AF522336

Fundaments of Animal Diseases

Fundaments of Animal Diseases

Lakshmi Prasanna Jakka

RANDOM PUBLICATIONS
NEW DELHI (INDIA)

Fundaments of Animal Diseases

ISBN 978-93-5111-884-8

Published in 2016 in India by

RANDOM PUBLICATIONS

4376-A/4B, Gali Murari Lal, Ansari Road
New Delhi-110 002
Phone : +9111-43580356, 011-23289044, 011-43142548
e-mail: sales@randompublications.com,
info@randompublications.com, randomexports@gmail.com
Reprinted 2019

Type Setting by : Friends Media, Delhi-110089
Digitally Printed at : Replika Press Pvt. Ltd.

Preface

Animal diseases generate a wide range of biophysical and socio-economic impacts that may be both direct and indirect, and may vary from very localised to global problems.

Concern with diseases that afflict animals dates from the earliest human contacts with animals and is reflected in early views of religion and magic. Diseases of animals remain a concern principally because of the economic losses they cause and the possible transmission of the causative agents to humans. The branch of medicine called veterinary medicine deals with the study, prevention, and treatment of diseases not only in domesticated animals but also in wild animals and in animals used in scientific research. The prevention, control, and eradication of diseases of economically important animals are agricultural concerns.

In developing countries, animal health services were established with the main objective of controlling major contagious and infectious diseases, such as foot-and-mouth disease, rinderpest and contagious pleuropneumonia, as well as parasitic diseases, such as trypanosomiasis and tick-borne diseases. This was obviously the first priority, since the control of these diseases is a prerequisite to any successful livestock development programme.

Disease control programmes are often established with the aim of eventual eradication of agents at a country, zone or compartment level. While this approach is desirable, the needs of stakeholders may require a broader range of outcomes. For some diseases, eradication may not be practically or economically feasible and options for sustained mitigation of disease impacts may be needed. It is important to clearly describe the programme goals and these may range from simple mitigation of disease impacts to progressive control or eradication of the disease. These guidelines highlight the importance of economic assessment of disease intervention options in the design of programmes taking into consideration effectiveness, feasibility of implementation, as well as costs and benefits

Animal Diseases has been carefully compiled and edited to meet the long felt needs of increasingly large number of those who have to deal with the different aspects of human diseases in colleges, universities and research institutes. It provides a stimulating and important new view of interaction between animals and pathogens causing diseases. The objective is to introduce to students the essential principles for understanding various aspects of diseases.

– ***Author***

Contents

1

Introduction

CONGENITAL DEFECTS

The cause of many congenital defects is unknown, but some are inherited. The most common inheritance patternis as a simple recessive trait. The defective calfreceives a recessive gene from its sire and one from itsdam. A few congenital defects are known to be causedby genes with incomplete dominance and a few arecaused by two or more sets of genes. Genetically caused congenital defects usually run infamilies. The parents of a genetically defective calf willgenerally have at least one ancestor in common. Whenmore than one genetically caused defective calf is bornin a herd in the same calving season, their dams areusually related (for example, half sisters) and are siredby the same bull. A change in the breeding programme isrequired to correct this situation.

Many congenital defects are caused by environmental factors. These include the level of nutrition, excess orshortages of certain nutrients, toxic plants or other toxic substances, infectious diseases, and extremes in temperature during pregnancy. Most environmentally caused congenital defects will occur during a short period of the calving season, from cows that were managed as a group. After proper diagnosis, a change in management is necessary to correct these conditions.

DIAGNOSING THE CAUSE

The cause of defects, the breeder must have good records and know why every calf dies. Breeding records which include sire and dam of each calf and breeding date are needed. Blood typing or DNA typing of the calf and possible parents can be used to help determine parentage. The calf must be alive and atleast one month old when the blood sample is obtained for blood typing. Management records should include which cows were in groups during each time period. Most breeders have a list of which cows are in each pasture. A date in and out of the pasture usually will help identify problems. Feed analysis reports, toxic plants present, and herd health and vaccination programmes are also of value. Meaningful the cause of death is important in controlling diseases as well as congenital defects. The cause of some deaths will be obvious, others will be much more difficult. If the breeder does not know the cause.

Genetic Defects

Dwarfism. There are several types of dwarfism caused by both environment and genetics. Each of the three genetically caused dwarfisms discussed here are different traits caused by different sets ofgenes.

- Snorter dwarfism causes short, blocky appearance with deformed bone growth in the nasal passages which causes difficulty in breathing. Inherited as a simple recessive trait.
- Long head dwarfism causes small size but does not affect the bone growth in nasal passages. Inherited as a simple recessive trait.
- Compress dwarfism is inherited as incomplete dominance. An individual with one compress gene and one normal gene has an extremely compressed body conformation. The individual with two compress genes is a dwarf and the calf dies at or soon after birth.

Water head (internal hydrocephalus). Excess fluid is present in the brain. Calves are usually born dead or die shortly after birth. Environmental factors can cause the disease, as well as being inherited as a simple recessive trait. Marble bone (osteopetrosis). The calves are usually born dead two to four weeks early. Bones are solid and do not contain marrow, making them very brittle and easily broken. Inherited as a simple recessive trait.

Hairlessness (hypotrichosis). Partial to almost complete lack of hair. Hair develops and is lost so an affected animal will vary somewhat in expression from month to month. Inherited as a simple recessive trait.

Rigid joints (arthrogryposis). Many environmentally caused forms appear but one form is inherited as a simple recessive trait. The joints of all four legs are fixed symmetrically and a cleft palate is present.

Extra toes (polydactyly). One or both front feet are usually affected, but all four may have the outerdew claw develop into an extra toe. Atleast two sets of genes are involved in the inheritance of this trait.

Mulefoot (syndactyly). The two toes are fused together to produce only one toe. The front feet are most often affected, but all four may be affected. These cattle cannot tolerate hot temperatures. Inherited as a simple recessive trait.

Weaver calf (progressive bovine my eloencephalopathy). Calves start developing a weaving gait at 6-8 months and get progressively worse until death at 12-20 months. Inherited as a simple recessive trait.

Photosensitivity (protoporphyria). Animals are sensitive to sunlight and develop scabs and open sores when exposed to sunlight. The liver is also affected and the animals may suffer from seizures. Inherited as a simple recessive trait.

Bulldog (achondroplasia). This trait is inherited as an incomplete dominant.

The homozygous may be aborted dead at 6-8 months gestation, and has acompressed skull, nose divided by furrows and shortened upper jaw, giving the bulldog facial appearance. The heterozygous calf is small and heavy-muscled.

Double muscling. Animals are extremely heavily muscled. However, considerable variation exists in the expression of this trait. Inherited as a simple recessive trait.

Parrot mouth (brachygnathia inferior). One type of parrot mouth is inherited as a simple recessive trait. The more common cause of teeth and denture pads not meeting is a quantitative trait caused by several sets of genes. This can cause either an under or over shot jaw with varying degrees of expression.

Cryptorchidism. One or both testicles fail to descend into the scrotum. Inherited as a sex limited trait and probably involves at least two sets of genes.

Prolonged gestation. The fetus fails to trigger parturition. Parturition must be induced or the calf removed. The calf is often extremely large and of tendies. Inherited as a simple recessive trait.

White eyes (Oculocutaneous Hypopigmentation). Hair coat is a bleached colour and the iris is pale blue around the pupil with tan periphery. Inherited as a simple recessive trait.

Many other genetically caused undesirable traits are known. The beef cattle geneticist at your land grant university will have knowledge of most of them and will beable to help if a problem arises.

Many abnormal conditions are not genetically caused. Two headed calves and calves with extra legsare caused by mistakes in development and not thegenetic makeup of the individual or its parents. Freemartin heifers are caused by circulation of the malet win's hormones through the developing female fetus. Some hydrocephalus can be caused by BVD (bovinevirus diarrhea) infection during pregnancy. Crippled-calfdisease is caused by the cow eating lupines between days 40 and 60 of pregnancy. Flexed pasterns (contractedflexor tendons) is usually caused by a large fetus developing in a small uterus. However, both crippled calf and flexed pasterns can also be genetically caused, inherited as simple recessive traits.

CARDIOVASCULAR DISEASES IN ANIMALS

When the incidences and types of cardiovascular disease present in various species, breeds, and strains, or families of animals are compared, certain differences are apparent. It is often difficult, however, to separate hereditary from environmental influences. Studies of vascular disease in zoo animals have shown that changing environmental conditions can alter the incidence of certain types of lesions in various species. Species differences, however, in resistance to diet-induced atherosclerosis appear to be genetically determined. The prevalence and types of congenital cardiac malformations appear to differ from species to species, but further systematic study is required. Arterial blood pressure is higher in the giraffe and turkey than in other species, and normal variants in cardiac rhythm are characteristic of the dog, horse, and mole.

Relatively high incidences of specific cardiovascular diseases are found in certain breeds of animals. The White Carneaux, Autosexing King, and Silver King breeds of pigeons have a high incidence of atherosclerosis. Congenital heart disease appears to be more common in purebred than in mongrel dogs, and an unusual aggregation of cases of subaortic stenosis in the Boxer and German Shepherd breeds has been found. In a survey of heart disease in dogs, the prevalence of chronic congestive heart failure in the male Cocker Spaniel greatly exceeded that in the male and female of all other breeds. Arterial blood pressure is higher in Broad Breasted Bronze turkeys than in the Jersey Buff breed. This is associated in the former breed with a relatively high incidence of spontaneous aortic rupture.

The occurrence of cardiovascular disease is unusually high in certain families and strains of animals. Among swine, litter and strain differences in serum cholesterol levels and in susceptibility to atherogenic diets occur. The White Carneaux breed of pigeons is actually a highly inbred strain with a remarkable predisposition to the development of atherosclerosis. Strains of chickens and rats with high incidences of interventricular septal defects have been developed by selective breeding. The familial occurrence of congenital heart disease in dogs and swine has been observed, and an inherited vascular anomaly in cattle has been described. Through selective breeding of laboratory rodents, strains with various types of cardiomyopathies have been developed. Certain diseases thought to be similar to the heritable disorders of connective tissue in man have been identified in domestic species. The level of arterial blood pressure is a heritable characteristic, and strains of rabbits, rats, and chickens with relative hypertension have been produced by selective breeding. Many of these observations indicate the importance of inheritance in determining susceptibility to various types of acquired cardiovascular disease. Genetic factors appear to operate in determining the occurrence of certain congenital malformations. Breeding experiments provide the most convincing evidence of genetic influence on the development of specific cardiovascular lesions. This experimental approach holds the greatest promise for furthering knowledge and understanding of the role of inheritance in the etiology of disease of the heart and blood vessels.

FUNCTION OF THE CARDIOVASCULAR SYSTEM

By circulating blood throughout the body, the cardiovascular system functions to supply the tissues with oxygen and nutrients, while removing carbon dioxide and other metabolic wastes. As oxygen-rich blood from the heart flows to the tissues of the body, oxygen and other chemicals move out of the blood and into the fluid surrounding the cells of the body's tissues. Waste products and carbon dioxide move into the blood to be carried away. As blood circulates through organs such as the liver and kidneys some of these waste products are removed. Blood then returns to the lungs (or gills, in the case of fish), receives a fresh dose of oxygen and gives off carbon dioxide. Then the cycle repeats itself. This process of circulation is necessary for continued life of the cells, tissues, and ultimately the whole organism. Up and down the evolutionary ladder, there are different forms of cardiovascular systems with different levels of efficiency, but they all perform this same basic function.

MAMMALIAN ANATOMY AND PHYSIOLOGY

The cardiovascular systems of mammals, birds, amphibians, reptiles, and fish are all slightly different. The following is an overview of the main components of the mammalian system – the heart and blood vessels.

Heart

The heart is composed of cardiac muscle that differs slightly from the skeletal and smooth muscle found elsewhere in the body. This special type of muscle adjusts the rate of muscular contraction, allowing the heart to maintain a regular pumping rhythm. The main parts of the heart are the chambers, the valves, and the electrical nodes.

Heart Chambers: There are two different types of heart chambers. The first is the atrium (plural is atria), which receives blood returning to the heart through the veins. The right atrium pumps blood to the right ventricle, and the left atrium pumps blood into the left ventricle. This blood is then pumped from the atrium into the second chamber called the ventricle. The ventricles are much larger than the atria and their thick, muscular walls are used to forcefully pump the blood from the heart to the body and lungs (or gills).

Valves: The valves found within the heart are situated between the atria and ventricles, and also between the ventricles and major arteries. These valves are opened and closed by pressure changes within the chambers, and act as a barrier to prevent the backflow of blood. The characteristic *'lub-dub, lub-dub'* heart sounds heard through a stethoscope are the result of vibrations caused by the closing of the respective valves.

Electrical Nodes: There are two different electrical nodes, or groups of specialized cells, located in the cardiac tissue. The first is the sinoatrial (SA) node, commonly called the pacemaker. The pacemaker is embedded in the wall of the right atrium. This small patch of tissue experiences rhythmic excitation and the impulse rapidly spreads throughout the atria, causing a muscular contraction and the pumping of blood from the atria to the ventricles. The other node, the atrioventricular (AV) node, relays the impulse of the SA node to the ventricles. It delays the impulse to prevent the ventricles from contracting at the same time as the atria, thus giving them time to fill with blood. The cycle of contraction of the heart muscle is called a heartbeat, the rate of which varies greatly between organisms. The following table gives the average heart rates of some common mammals.

Table 4.1 : Heart Rates Comparison (beats/minute)

Organism	*Average Rate*	*Normal Range*
Human	70	58-104
Cat	120	110-140
Cow	65	60-70
Dog	115	100-130
Guinea Pig	280	260-400
Hamster	450	300-600
Horse	44	23-70
Rabbit	205	123-304
Rat	328	261-600

Vessels

A vessel is a hollow tube for transporting something, like a garden hose transporting water. A blood vessel is a hollow tube for transporting blood. There are three main types of blood vessels:

- Arteries
- Capillaries
- Veins.

These main blood vessels function to transport blood through the entire body and exchange oxygen and nutrients for carbon dioxide and wastes. The arteries carry blood away from the heart, and are under high pressure from the pumping of the heart.

To maintain their structure under this pressure, they have thick, elastic walls to allow stretch and recoil. The large pulmonary artery carries unoxygenated blood from the right ventricles to the lung, where it gives off carbon dioxide and receives oxygen. The aorta is the largest artery. It carries oxygenated blood from the left ventricle to the body. The arteries branch and eventually lead to capillary beds. The capillaries make up a network of tiny vessels with extremely thin, highly permeable walls. They are present in all of the major tissues of the body and function in the exchange of gases, nutrients, and fluids between the blood, body tissues, and alveoli of the lungs.

At the opposite side of the capillary beds, the capillaries merge to form veins, which return the blood back to the heart. The veins are under much less pressure than the arteries and therefore have much thinner walls.

The veins also contain one-way valves in order to prevent the blood from flowing the wrong direction in the absence of pressure. The pulmonary vein returns oxygenated blood from the lungs to the left atria. The vena cava returns blood from the body to the right atria. The blood that is returned to the heart is then recycled through the cardiovascular system.

COMPARATIVE ANATOMY

Mammals and Birds

Mammalian and avian hearts have four chambers – two atria and two ventricles. This is the most efficient system, as deoxygenated and oxygenated bloods are not mixed. The right atrium receives deoxygenated blood from the body through both the inferior and superior vena cava. The blood then passes to the right ventricle to be pumped through the pulmonary arteries to the lungs, where it becomes oxygenated.

It returns to the left atrium via the pulmonary veins, this oxygen-rich blood is then passed to the left ventricle and pumped through the aorta to the rest of the body.

The aorta is the largest artery and has an enormous amount of stretch and elasticity to withstand the pressure created by the pumping ventricle. The four-chambered heart ensures that the tissues of the body are supplied with oxygen-saturated blood to facilitate sustained muscle movement. Also, the larger oxygen supply allows these warm-blooded organisms to achieve thermoregulation (body temperature maintenance).

Amphibians and Reptiles

Amphibians and reptiles, by contrast, have a three-chambered heart. The three-chambered heart consists of two atria and one ventricle. (The crocodile is sometimes said to have a four-chambered heart. The separation of the ventricles is not complete, however, because a hole remains in the septum (wall) that divides the two chambers.) Blood leaving the ventricle passes into one of two vessels.

It either travels through the pulmonary arteries leading to the lungs or through a forked aorta leading to the rest of the body. Oxygenated blood returning to the heart from the lungs through the pulmonary vein passes into the left atrium, while deoxygenated blood returning from the body through the sinus venosus passes into the right atrium. Both atria empty into the single ventricle, mixing the oxygen-rich blood returning from the lungs with the oxygen-depleted blood from the body tissues. While this system assures that some blood always passes to the lungs and then back to the heart, the mixing of blood in the single ventricle means the organs are not getting blood saturated with oxygen. This is not as efficient as a four-chambered system, which keeps the two circuits separate, but it is sufficient for these cold-blooded organisms.

The heart rate of amphibians and reptiles is very dependent upon temperature. For example, the following table gives the approximate heart rate of a crocodile at the indicated temperatures. Notice that the higher the temperature, the faster the heart beat.

Table 4.2

Temperature (Celsius)	*Average Rate (beats/minute)*
10 C	1-8
18 C	15-20
28 C	24-40
>40 C	Irreversible cardiac damage

Fish

Fish possess the simplest type of true heart – a two-chambered organ composed of one atrium and one ventricle. A rudimentary valve is located between the two chambers. Blood is pumped from the ventricle through the conus arteriosus to the gills. The conus arteriosus is like the aorta in other species.

At the gills, the blood receives oxygen and gets rid of carbon dioxide. Blood then moves on to the organs of the body, where nutrients, gases, and wastes are exchanged. There is no division of the circulation between the gills and the body. That is, the blood travels from the heart to the gills, and then directly to the body before returning to the atrium through the sinus venosus to be circulated again.

The heart rates of fish fall within the wide range of 60-240 beats per minute, depending upon species and water temperature. The fish's heart rate will be slower at lower temperatures.

DIAGNOSIS OF THE BOVINE GENITAL TRACT DISORDERS

In bovine practice, ultrasonography has become an important diagnostic tool for evaluating the female reproductive system. Its importance in diagnosis lies in the non-invasiveness of the instrument.

In spite of its immense use in supplementing diagnosis during physiological states, its use in delineating different pathological conditions of the bovine genital tract continues to be less frequently described.

This study was conducted on clinical cases presented to the veterinary gynaecology and obstetrics outdoor to record the sonographic appearance during different pathological conditions of genital tract.

Materials And Methods

Cows presented to the veterinary gynaecology and obstetrics outdoor (n=85) with different pathological conditions were included in the present study. Transrectal sonographic examination was performed using a portable ultrasound machine (AGROSCAN linear, ECM 1"6 BD de la Republique, F 16000 Angouleme, FRANCE), with a linear array dual frequency probe (5.0/7.5 MHz). The images were saved in a multimedia kit attached to the instrument and subsequently transferred to the computer. Animals were examined by rectogenital palpation and ultrasonography was done later to confirm/potentiate the clinical diagnosis. After administration of an appropriate treatment sonography was done again to determine the effect of treatment.

CONTROLLING GENETIC DISEASES

The greatest control of genetic diseases is to avoid animals that carry these genes. Bulls or semen should be purchased from reputable breeders, produced by parents who are not known to carry undesirable genes. Long established inbred lines that have not recently produced genetic undesirables are usually quite safe. Commercial producers who use a crossbreeding system rarely have a problem. The elite purebred breeder or owner of Al bulls may wish to test for simply inherited traits before bulls or donor cows are heavily used. If the undesirable trait is dominant, no test is needed since the animal would show the trait even if only one dominant gene is present. If the trait is inherited as incomplete dominance, the individual that has only one undesired gene can usually be identified and testing is not needed. Testing is usually useful only when the trait is inherited as a simple recessive trait.

The slightest expensive test is to mate the animals to ones having that undesired trait. For example, if horns are not desired, the polled animal to be tested would be mated to horned animals to produce at least 7 calves (P>.99). If any horned calves are produced, the polled animal has one gene for horns. If the undesired trait is lethal, the dead animals cannot be used to make the test. If only one lethal trait is of concern, the test should be made using animals that have produced calves with this lethal trait. At least 16 calves should be produced from these matings (P>.99). If a calf is produced that has this lethal trait, the tested animal has one recessive gene for the trait. If the breeder is concerned about identifying all recessive traits, at least 36 progeny from sire-daughter or mother-son matings should be produced (P>.99). This test is time consuming and expensive and only truly outstanding animals will be able to pay for it. All calves from these matings must be observed very closely and all recessive traits recorded. The tested animal will have a recessive gene for any recessive traits observed.

What do Do with Carriers

The animal's desirable genes should be weighed against its undesirable genes. If the desirable genes can be found in other animals without the undesirable gene, carriers should be slaughtered and replaced. When the production traits are superior, these animals can be used in a cross breeding programme to produce beef. Heifers should not be kept for breeding. If the individual is extremely superior in production traits, a superior son can be produced that does not carry the undesirable gene. The outstanding carrier animal would be mated to a small group of very outstanding individuals. The best two to four sons produced would be selected and used in test matings to known carrier cows. The best son that does not carry the undesired gene would then be used and all carriers slaughtered. This would take several years and only truly superior individuals could justify such a procedure. In most cases, the animal that carries the undesirable recessive gene should not be used to produce breeding animals. Daughters should be worked out of the herd and replaced with superior animals that do not carry undesirable genes. Purebred breeders should work with their breed associations, extension and university personnel, and veterinarian to eliminate and avoid problems.

Ethical and Legal Considerations

Serious ethical and legal problems are involved in selling known carrier cattle or progeny of known carriers. A seed stock producer in this position should be completely honest with the buyer. It is doubtful that he should sell possible carriers, under any circumstances, to a youngster or to someone who is just getting started in business and may not have the knowledge to understand the consequences of using offspring from known carriers. Selling carriers without informing the buyer will ultimately reduce the confidence that buyers have in the breeder and may eventually reflect negatively on the entire breed.

GENETIC ABNORMALITIES IN BEEF CATTLE

Hereditary defects can cause abortion or be present at time of birth. They are uncommon but do occur in most breeds of cattle. Defects are abnormalities in skeleton, body form, and body functions. Abnormalities may result from genetic or environmental causes. When the environment is the cause, adjustments can reduce further economic losses. However, genetic (inherited) causes are much more complex and difficult to correct.

Environmental Causes

Environmental or non-genetic causes have the same economic results as genetic causes but are far easier to rectify. Simply correcting the environment will remove the problem. There are many environmental factors, including disease and diet. Certain conditions show that an abnormality is likely to be environmental in nature:

- The abnormality coincided with an environmental factor and was absent upon removal of the factor.
- The abnormality occurred in groups of non-related individuals.
- The symptoms are similar to those of an abnormality known to result from environmental factors.

Genetic Causes

Chromosomes inherited from parents determine an animal's genetic make-up. There are many genes in each chromosome. Genetic abnormalities occur when genes are missing, in excess, mutated or in the wrong location (translocation). A few genes can directly cause an abnormality, however, these are rare. Usually, these genes are recessive, meaning two must be present to cause an abnormality. Both parents must be carriers of the gene for a calf to be abnormal. In this case, only one of every four offspring will be abnormal. Two will be carriers and one will be normal. Certain conditions show that an abnormality is likely to have a genetic origin:

- The abnormality is more common in a group of related animals.
- The symptoms are similar to those of an abnormality identified through test matings. Study of an animal's chromosomes using blood samples can identify several genetic defects.

COMMON GENETIC DEFECTS

Hypotrichosis (Hairlessness)

Hairlessness occurs in several breeds of beef cattle. It expresses itself as complete or partial loss of hair. Calves are often born with no hair but will grow a short curly coat of hair with age. Affected individuals are prone to environmental stress (cold and wet) and skin infections are more prevalent. A recessive gene causes hairlessness.

Alopecia Anemia

This syndrome has recently been identified in the Polled Hereford breed. At the time of birth, alopecia anaemia may be mistaken for hairlessness. Affected calves are often small at birth, have a dirty-faced appearance, and have protruding tongue and eyes.

Hair is wiry, tightly curled or absent while wrinkled skin gives the appearance of advanced aging. Calves are lethargic, cannot tolerate stress and are very prone to disease. Few survive past six months of age. Malfunction of the skeletal structure results in reduced red blood cell production (anaemia). Alopecia anaemia occurs in families but the exact mode of transmission is unknown.

Translocations

A translocation occurs when part of a chromosome breaks off and attaches to another chromosome. The 1/29 translocation has been identified in the Simmental, Charolais and Blonde D'Aquitaine breeds. The 14/20 translocation occurs in most Continental breeds. Translocations affect fertility but no other production traits. Carriers of translocations have reduced conception rates and increased abortion rates. Blood analysis allows easy identification of carriers.

Beta-mannosidosis (Beta-man)

The Beta-man disorder is due to a recessive gene that produces a defective enzyme. The result is the birth of calves that never get up and eventually die. The syndrome occurs in the Salers breed and a blood test is available for identifying carriers.

Syndactyly (Mulefoot)

Syndactyly refers to the fusion of the two toes of the foot. Caused by a recessive gene, mulefoot most often affects the front feet. This condition occurs in the Aberdeen Angus breed.

Other genetic defects exist, most being of very low frequency.

What Should You Do?

When you suspect that you have a problem calf, consult your veterinarian and OMAFRA extension specialist. Investigate all symptoms and possible causes before concluding the problem is genetic or environmental. When the cause is genetic, contact the breed association and give them a full report of the findings. Progressive breed associations are working to reduce the frequency of genetic abnormalities within their breed. To avoid further abnormalities in your herd without culling female carriers, use non-carrier bulls unrelated to your herd. Practice no inbreeding within the herd. Crossbreeding to a different breed is another alternative.

HEREDITARY DISEASE

Congenital erythropoietic porphyria is a rare hereditary disease of cattle, pigs, cats, and humans in which defective hemoglobin formation results in production of an excess of Type I porphyrins in the nuclei of developing normoblasts. The defect in cattle is inherited as a simple autosomal recessive and is usually confined to herds in which inbreeding or close line-breeding is practiced. The condition has been recognized in the USA, Canada, Denmark, Jamaica, England, South Africa, Australia, and Argentina. This broad geographic distribution indicates that the disease likely occurs worldwide and probably affects all meat-producing animals, especially cattle, swine, and sheep.

Heterozygous animals seem to be normal, but homozygous recessive animals are affected at birth with reddish brown discoloration of the teeth, bones, and urine that persists for the life of the animal. The inherited enzymatic defect causes deficient activity of uroporphyrinogen III synthase—an essential part of porphyrin-heme biosynthesis. Uroporphyrinogen III cosynthetase is the enzyme that is deficient. The urine contains an excess of coproporphyrin I and uroporphyrin I; in affected animals, the colour is amber or reddish brown. Bones, urine, and teeth (especially the deciduous teeth) fluoresce pink when irradiated with near-ultraviolet light. Prolonged exposure to sunlight causes typical lesions of photosensitization with hyperemia, vesicle formation, and superficial necrosis of unpigmented portions of the skin. The severity of the skin lesions depends on the intensity of the solar radiation and the extent of cutaneous pigmentation occurring in specific families of animals. A normochromic, hemolytic anemia with macrocytes and microcytes and marked basophilic stippling develops. Splenomegaly eventually occurs. The texture of bones is not altered except in cases in which bones have increased fragility due to a diminished cortex. Affected animals are generally of medium to good condition unless solar injury has occurred. Some animals become progressively unthrifty unless protected from sunlight. A similar disease, bovine protoporphyria, causes photosensitivity only in Limousin cattle and humans.

In humans, a series of porphyrias caused by defective functions of enzymes in porphyrin-heme biosynthesis have been described and grouped according to their presenting clinical signs. These vary broadly and may include severe cutaneous lesions on exposed areas of the body, acute photosensitivity reactions, serious liver damage, and acute attacks of neurologic dysfunction. In animals, the recognized diseases are commonly classified as either congenital erythropoietic porphyria, congenital erythropoietic protoporphyria, or porphyria. It is likely that all of the syndromes described in humans also occur in animals and that a broader classification could be used.

The defect in pigs and cats is extremely rare and differs from the condition in cattle in that photosensitization is not a feature. In pigs and cats, the disease is transmitted as an autosomal dominant. In pigs, even with high levels of porphyrins in the blood, photodynamic dermatitis does not occur. The disease has been reported only in Denmark and New Zealand; in cats, it has been recognized only in the USA. Diagnosis should be based on the excretion of abnormal uroporphyrins, the brown discoloration of the teeth (which fluoresce when irradiated with near-ultraviolet light), the appearance of discoloured urine, and hemolytic anemia.

The recessive genetic character is widely distributed in cattle, but the clinical condition is comparatively rare. Clinically normal heterozygotes have lower levels of uroporphyrinogen III cosynthetase than do normal animals, but laboratory identification of the carrier state is impractical due to the relatively low incidence of the disease and is not widely used. Morbidity can be controlled by keeping affected animals indoors and out of direct sunlight.

INTEGUMENTARY SYSTEM

Skin

Dermatitis is an inflammation of the skin. A bacterial infection is one cause of dermatitis. During warm, humid weather, the skin and haircoat is more susceptible to bacterial infection and the prevalence of skin diseases is highest. The most common sign is pruritus (itching). This can be followed by skin lesions that progress from reddening and thickening of the skin to bumps, blisters, and crusts or scales. Other causes of dermatitis include viruses, parasites, fungi, and allergens. Pyoderma is dermatitis characterized by the presence of purulent exudate (pus). Dermatitis can develop into pyoderma due to the invasion of pus-forming bacteria. These infections can be superficial or deep pustules with draining tracts. Bacterial dermatitis and pyoderma are commonly associated with skin allergy and mange conditions.

Dermatophytosis, or ringworm, is a result of a pathogenic fungus. This fungus infects the skin of dogs, cats, horses, cattle, pigs, humans, sheep, and goats. The normal habitat is in the skin but can survive in dark humid environments of soil, bedding, carpet, furniture, tack, blankets, brushes, and clippers. Transmission of ringworm occurs through direct contact with infected animals and humans or contaminated environment. The fungus is transmitted through skin contact from a carrier dam to her nursing young. The most susceptible hosts appear to be the young and they often display the characteristic lesions. The fungal infection may not be evident until several months after exposure. The cutaneous disease begins as focal alopecia (hair loss). These round areas of hair loss become scaly with circumscribed edges that may be raised and reddened. Pruritus may or may not be present. Use direct microscopic examination of skin scrapings and skin fluorescent examinations with a ultraviolet lamp (Wood's lamp) to make ringworm diagnosis. Confirm the diagnosis using culture techniques. Since spontaneous recovery after several months is common, the primary objective of topical or oral therapy is to prevent spread of the lesions and spread of infection to other animals and humans.

Dermatophilosis, or rain gall, is a result of a fungus (Dermatophilus spp.). This disease is common in horses, cattle, sheep, and goats that live in warm, humid, damp climates. During the rainy season, biting flies may help transmit the disease from infected animals and from wet contaminated soils by acting as a mechanical vector. The cutaneous disease is an exudative dermatitis with scabs. Typical lesions are raised, crusty lumps covered with hair that can be pulled off. Removal of the crust with a tuft of hair (paintbrush lesions) leaves a bare spot. These lesions commonly develop on the lower legs, chest, back, and hips. Use cultures, biopsies, and scrapings to make a definitive diagnosis of this fungal disease. Warts are fibrous tumors of the skin and occasionally of the mucous membranes of animals (especially cattle, dogs, rabbits) and humans. They can be caused by many strains of viruses. The virus is transmitted by direct contact and possibly by arthropods. The cauliflower-type growths occur primarily on the head, neck, and shoulder, in the mouth, and on the vulva and penis.

Pox diseases cause skin lesions by replication through poxviruses in the skin. Animals acquire the virus through the skin or by biting arthropods. Bumps, blisters, pustules, and crusts are the types of skin lesions present in the course of the disease. The pox diseases are named after the affected animals, such as fowlpox, swinepox, and cowpox.

Arthropod parasites are external parasites, or ecto-parasites. Their presence, or the dermatitis they cause, is referred to as an infestation instead of an infection. Therefore, external parasites infest an animal. Parasitic infestations of the skin affect the health of animals by causing tissue damage, blood loss, and annoyance. Annoying pests interfere with the animal's ability to eat and sleep, which may result in weight loss. In addition to the skin diseases directly caused by arthropods, many ectoparasites are vectors of infectious diseases, transmitting disease agents from carrier animals to other susceptible animals.

Some flies, such as houseflies, do not suck blood but annoy animals. Other flies, such as horseflies, deer-flies, stableflies, and hornflies, are blood-suckers and cause anemia, as well as, annoyance. Blood-sucking flies, mosquitoes, and gnats aid the transmission of diseases, such as anaplasmosis, bluetongue, leukosis, equine infectious anemia, and heartworms.

Heelflies have larval stages, called cattle grubs that migrate through the body and emerge through the skin on the backs of cattle. Large numbers of cattle grubs may cause migratory damage to internal tissues and the hide. Ticks are subdivided into two groups: hard ticks and soft ticks. The hard ticks attach and feed on animals by sucking blood for several days; many types serve as vectors of diseases. Hard ticks help transmit anaplasmosis and Lyme disease to other animals. Soft ticks common in fowl are intermittent feeders. The immature stages of spinose ear tick, a soft tick, develop in the external ear canal of animals, especially cattle, but the adult ticks are free-living.

Dogs and cats are commonly infested with fleas. Fleas are blood-suckers and can be vectors of the tapeworm, called the flea tapeworm. Fleas are extremely annoying to pets and may cause severe anemia. Lice infestations of animals are more common during the cool and cold months of the year, especially in late winter and early spring. During this time animals are in close contact with one another, their skin has less oil, and lice reproduction is greater. Lice are more common on cattle, swine, and poultry than on other animals. Both biting lice and sucking lice annoy animals and cause skin allergies and hair loss. In large numbers, the sucking lice can cause anemia. Lousy is the condition of an animal with a lice infestation. Lice and their eggs are easily visible on animals with the naked eye.

The entire life cycle of mange mites is completed on or in the skin and ear mites in external ear canals of animals. Transmission occurs through direct contact. Mites infesting the skin of animals cause a condition called mange. Mange is more common in the winter and during times of stress. Mange is common in dogs and swine, occasionally in cattle, and rarely in other animals. Types of mange conditions with hair loss and dermatitis are sarcoptic mange, scabies, and red mange. Ear mites common in dogs and cats do not cause mange but do cause extreme annoyance. Microscopic examination of skin scrapings from infested animals reveals the presence of surface mites or burrowing mites and ear swabs the presence of ear mites.

Nematodes, or roundworms, infect the skin of animals. Examples of nematodes include habronema, on-chocerca, stephanofilaria, and dipetalonema. House-flies and stableflies help transmit habronema larvae that infect wounds and external mucous membranes of horses. They can cause excessive granulation (proud flesh), a condition called summer sores. Onchocerca larvae, transmitted by mosquitoes and gnats, can cause an allergic skin reaction on the face, neck, chest, and underline of horses. Hornflies transmit stephanofilaria to cattle and cause local circumscribed lesions on the underline. Fleas transmit dipetalonema to the skin of dogs but it causes no harm.

THE INTEGUMENTARY SYSTEM

The integumentary system, formed by the skin, hair, nails, and associated glands, enwraps the body. It is the most visible organ system and one of the most complex. Diverse in both form and function—from delicate eyelashes to the thick skin of the soles—the integumentary system protects the body from the outside world and its many harmful substances. It utilizes the Sun's rays while at the same time shielding the body from their damaging effects. In addition, the system helps to regulate body temperature, serves as a minor excretory organ, and makes the inner body aware of its outer environment through sensory receptors.

Design: Parts of the Integumentary System

Integument comes from the Latin word *integumentum*, meaning 'cover' or 'enclosure.' In animals and plants, an integument is any natural outer covering, such as skin, shell, membrane, or husk. The human integumentary system is an external body covering, but also much more. It protects, nourishes, insulates, and cushions. It is absolutely essential to life. Without it, an individual would be attacked immediately by bacteria and die from heat and water loss. The integumentary system is composed primarily of the skin and accessory structures. Those structures include hair, nails, and certain exocrine glands (glands that have ducts or tubes that carry their secretions to the surface of the skin or into body cavities for elimination).

Skin

Although the skin is not often thought of as an organ, such as the heart or liver, medically it is. An organ is any part of the body formed of two or more tissues that performs a specialized function. As an organ, the skin is the largest and heaviest in the body. In an average adult, the skin covers about 21.5 square feet (2 square meters) and accounts for approximately 7 percent of body weight, or about 11 pounds (5 kilograms) in a 160-pound (73-kilogram) person. It ranges in thickness from 0.04 to 0.08 inches (1 to 2 millimetres), but can measure up to 0.2 inches (6 millimetres) thick on the palms of the hands and the soles of the feet. The skin in these areas is referred to as thick skin (skin elsewhere on the body is called thin skin).

The Integumentary System: Words to Know

Apocrine sweat glands (AP-oh-krin): Sweat glands located primarily in the armpit and genital areas.

Arrector pili muscle (ah-REK-tor PI-li): Smooth muscle attached to a hair follicle that, when stimulated, pulls on the follicle, causing the hair shaft to stand upright.

Dermal papillae (DER-mal pah-PILL-ee): Finger-like projections extending upward from the dermis containing blood capillaries, which provide nutrients for the lower layer of the epidermis; also form the characteristic ridges on the skin surface of the hands (fingerprints) and feet.

Dermis (DER-miss): Thick, inner layer of the skin.

Eccrine sweat glands (ECK-rin): Body's most numerous sweat glands, which produce watery sweat to maintain normal body temperature.

Epidermis (ep-i-DER-miss): Thin, outer layer of the skin.

Epithelial tissue (ep-i-THEE-lee-al): Tissue that covers the internal and external surfaces of the body and also forms glandular organs.

Integument (in-TEG-ye-ment): In animals and plants, any natural outer covering, such as skin, shell, membrane, or husk.

Keratin (KER-ah-tin): Tough, fibrous, water-resistant protein that forms the outer layers of hair, calluses, and nails and coats the surface of the skin.

Lunula (LOO-noo-la): White, crescent-shaped area of the nail bed near the nail root.

Melanocyte (MEL-ah-no-site): Cell found in the lower epidermis that produces the protein pigment melanin.

Organ (OR-gan): Any part of the body formed of two or more tissues that performs a specialized function.

Sebaceous gland (suh-BAY-shus): Exocrine gland in the dermis that produces sebum.

Sebum (SEE-bum): Mixture of oily substances and fragmented cells secreted by sebaceous glands.

Squamous cells (SKWA-mus): Cells that are flat and scalelike.

Subcutaneous (sub-kew-TAY-nee-us): Tissues between the dermis and the muscles.

The skin has two principal layers: the epidermis and the dermis. The epidermis is the thin, outer layer, and the dermis is the thicker, inner layer. Beneath the dermis lies the subcutaneous layer or hypodermis, which is composed of adipose or fatty tissue. Although not technically part of the skin, it does anchor the skin to the underlying muscles. It also contains the major blood vessels that supply the dermis and houses many white blood cells, which destroy foreign invaders that have entered the body through breaks in the skin.

Epidermis

The epidermis is complete of stratified squamous epithelial tissue. Epithelial tissue covers the internal and external surfaces of the body and also forms glandular organs. Squamous cells are thin and flat like fish scales. Stratified simply means having two or more layers. In short, the epidermis is composed of many layers of thin, flattened cells that fit closely together and are able to withstand a good deal of abuse or friction. The epidermis can be divided into four or five layers. Most important of these are the inner and outer layers. The inner or deepest cell layer is the only layer of the epidermis that receives nutrients (from the underlying dermis). The cells of this layer, called basal cells, are constantly dividing and creating new cells daily, which push the older cells towards the surface. Basal cells produce keratin, an extremely durable and water-resistant fibrous protein.

Another type of cell found in the lower epidermis is the melanocyte. Melanocytes produce melanin, a protein pigment that ranges in colour from yellow to brown to black. The amount of melanin produced determines skin colour, which is a hereditary characteristic. The melanocytes of dark-skinned individuals continuously produce large amounts of melanin. Those of light-skinned individuals produce less. Freckles are the result of melanin clumping in one spot. The outermost layer of the epidermis consists of about twenty to thirty rows of tightly joined flat dead cells. All that is left in these cells is their keratin, which makes this outer layer waterproof. It takes roughly fourteen days for cells to move from the inner layer of the epidermis to the outer layer. Once part of the outer layer, the dead cells remain for another fourteen days or so before flaking off slowly and steadily.

Dermis

The dermis, the second layer of skin, lies between the epidermis and the subcutaneous layer. Much thicker than the epidermis, the dermis contains the accessory skin structures. Hair, sweat glands, and sebaceous (oil) glands are all rooted in the dermis. This layer also contains blood vessels and nerve fibres. Nourished by the blood and oxygen provided by these blood vessels, the cells of the dermis are alive. Connective tissue forms the dermis. Bundles of elastic and collagen (tough fibrous protein) fibres blend into the connective tissue. These fibres provide the dermis strength and flexibility.

The upper layer of the dermis has fingerlike projections that extend into the epidermis. Called dermal papillae, they contain blood capillaries that provide nutrients for the basal cells in the epidermis. On the skin surface of the hands and feet, especially on the tips of the fingers, thumbs, and toes, the dermal papillae form looped and whorled ridges. These print patterns, known as fingerprints or toeprints, increase the gripping ability of the hands and feet. Genetically determined, the patterns are unique to every individual.

Using Fingerprints to Identify Community

Fingerprints are unique to each individual and the patterns never change. People have long known about the distinctiveness of fingerprints, but their use in identifying people did not arise until the nineteenth century. It is generally acknowledged that English scientist Francis Galton (1822–1911) was the first person to devise a system of fingerprint identification. In the 1880s, Galton obtained the first extensive collection of fingerprints for his studies on heredity. He also established a bureau for the registration of civilians by means of fingerprints and measurements.

Galton's ideas were further developed by fellow Englishman Edward R. Henry (1850–1931). In the 1890s, Henry developed a more simplified fingerprint classification system. In 1901, he established England's first fingerprint bureau, called the Fingerprint Branch, within the Scotland Yard police force. Henry's system is still used today in Great Britain and the United States. Within the dermis are sensory receptors for the senses of touch, pressure, heat, cold, and pain. A specific type of receptor exists for each sensation. For pain, the receptors are free nerve endings. For the other sensations, the receptors are encapsulated nerve endings, meaning they have a cellular structure around their endings. The number and type of sensory receptors present in a particular area of skin determines how sensitive that area is to a particular sensation. For example, fingertips have many touch receptors and are quite sensitive. The skin of the upper arm is less sensitive because it has very few touch receptors.

Accessory Structures

The accessory structures of the integumentary system include hair, nails, and sweat and sebaceous glands.

Hair

Roughly 5 million hairs cover the body of an average individual. About 100,000 of those hairs appear on the scalp. Almost every part of the body is covered by hair, except the palms of the hands, the soles of the feet, the sides of the fingers and toes, the lips, and certain parts of the outer genital organs.

Each hair originates from a tiny tubelike structure called a hair follicle that extends deep into the dermis layer. Often, the follicle will project into the subcutaneous layer. Capillaries and nerves attach to the base of the follicle, providing nutrients and sensory information. Inside the base of the follicle, epithelial cells grow and divide, forming the hair bulb or enlarged hair base.

Keratin, the primary component in these epithelial cells, coats and stiffens the hair as it grows upward through the follicle. The part of the hair enclosed in the follicle is called the hair root. Once the hair projects from the scalp or skin, it is called a hair shaft.

The older epithelial cells forming the hair root and hair shaft die as they are pushed upward from the nutrient-rich follicle base by newly formed cells. Like the upper layers of the epidermis, the hair shaft is made of dead material, almost entirely protein.

The hair shaft is divided into two layers: the cuticle or outer layer consists of a single layer of flat, overlapping cells; the cortex or inner layer is made mostly of keratin. Hair shafts differ in size, shape, and colour. In the eyebrows, they are short and stiff, but on the scalp they are longer and more flexible. Elsewhere on the body they are nearly invisible. Oval-shaped hair shafts produce wavy hair. Flat or ribbonlike hair shafts produce kinky or curly hair. Perfectly round hair shafts produce straight hair. The different types of melanin—yellow, rust, brown, and black—produced by melanocytes at the follicle base combine to create the many varieties of hair colour, from the palest blonde to the richest black. With age, the production of melanin decreases, and hair colour turns gray. Attached to each hair follicle is a ribbon of smooth muscle called an arrector pili muscle. When stimulated, the muscle contracts and pulls on the follicle, causing the hair shaft to stand upright.

Nails

Nails in humans correspond to the hooves of horses and cattle and the claws of birds and reptiles. Found on the ends of fingers and toes, nails are produced by nail follicles just as hair is produced by hair follicles. The nail root is that portion of the nail embedded in the skin, lying very near the bone of the fingertip. Here, cells produce a stronger form of keratin than is found in hair. As new cells are formed, older cells are pushed forward, forming the nail body or the visible attached portion of the nail. The free edge is that portion of the nail that extends over the tip of the finger or toe. Healthy fingernails grow about 0.04 inches (1 millimetre) per week, slightly faster than toenails. The nail body is made of dead cells, but the nail bed (the tissue underneath the nail body) is alive. The blood vessels running through the nail bed give the otherwise transparent nail body a pink colour. Near the nail root, however, these blood vessels are obscured. The resulting white crescent is called the lunula (from the Latin word *luna*, meaning 'moon').

Sweat Glands

More than 2.5 million sweat glands are distributed over most surfaces of the human body. They are divided into two types: eccrine sweat glands and apocrine sweat glands. Eccrine glands, the more numerous of the two types, are found all over the body. They are especially numerous on the forehead, upper lip, palms, and soles. The glands are simply coiled tubes that originate in the dermis. A duct extends from the gland to the skin's surface, where it opens into a pore. Eccrine glands produce sweat or perspiration, a clear secretion that is 99 percent water. Some salts, traces of waste materials such as urea, and vitamin C form the remainder (the salts give sweat its characteristic salty taste).

Depending on temperature and humidity, an average individual loses 0.6 to 1.7 quarts (0.3 to 0.8 liters) of water every day through sweating. During rigorous physical activity or on a hot day, that amount could rise to 5.3 to 7.4 quarts (5 to 7 liters).

Apocrine glands are found in the armpits, around the nipples, and in the groin. Like eccrine glands, apocrine glands are coiled tubes found in the dermis. However, they are usually larger and their ducts empty into hair follicles. Also, apocrine glands do not function until puberty. At that time, they begin to release an odourless cloudy secretion that contains fatty acids and protein. If the secretion of apocrine glands is allowed to remain on the skin for any length of time, bacteria that lives on the skin breaks down the fatty acids and protein for their growth, creating the unpleasant odour often associated with sweat. Apocrine glands are activated by nerve fibres during periods of pain and stress, but their function in humans is not well understood. Scientists theorize they may act as sexual attractants.

Sebaceous Glands

Sebaceous glands, also known as oil glands, are found in the dermis all over the body, except for the palms and soles. They secrete sebum, a mixture of lipids (fats), proteins, and fragments of dead fatproducing cells. The function of sebum is to prevent the drying of skin and hair. It also contains chemicals that kill bacteria present on the skin surface. While most sebaceous glands secrete sebum through ducts into hair follicles, some secrete sebum directly onto the surface of the skin. Arrector pili muscles, which contract to elevate hairs, also squeeze sebaceous glands, forcing out sebum.

FUNCTIONS OF INTEGUMENTARY SYSTEM

The integumentary system is essential to the body's homeostasis, or ability to maintain the internal balance of its functions regardless of outside conditions. The system works to protect underlying tissues and organs from infections and injury. It also prevents the loss of body fluids. Receiving about one-third of the blood pumped from the heart every minute, the skin and its glands help maintain normal body temperature.

The system also acts as a mini-excretory system, secreting salts, water, and wastes in the form of sweat. Cells in the skin utilize sunlight to create vitamin D, which is necessary for normal bone growth and function. Finally, the skin contains sensory receptors or specialized nerve endings that allow an individual to 'feel' sensations such as touch, pain, pressure, and temperature.

Protection

The outermost epidermal layer of the skin is a barrier between the internal environment of the body and the external world. Keratin, in abundance in this outer layer, waterproofs the body. Without it, handling household chemicals, swimming in a pool, or taking a shower (a necessary everyday activity) would be disastrous to the underlying cells of the body. Not only does keratin keep water out, it also keeps water in. Excessive evaporation or loss of body fluids would result in dehydration and eventual death.

WETTERHAHN'S DEADLY RESEARCH

Karen Wetterhahn (1948–1997) was a chemistry professor at Dartmouth College in Hanover, New Hampshire, where she conducted environmental research projects. During an experiment in August 1996, Wetterhahn spilled a tiny drop of dimethyl mercury (a highly toxic chemical) on her hand. Less than a year later, she was dead. Wetterhahn had been conducting research to determine the effects that heavy metals (metals such as mercury having a high specific gravity) produce on the environment. During her experiment, she was transferring some dimethyl mercury to a tube when she spilled a tiny amount. Although Wetterhahn was wearing latex gloves, the mercury permeated the thin latex and soaked into her skin, passing through its waterproof layers within seconds.

Dimethyl mercury is deadly. Once in the body, it seeps from the bloodstream into brain tissues, causing fatal damage to the central nervous system and the brain. Symptoms of mercury poisoning include loss of motor (movement) control, numbness in the arms and legs, blindness, hearing and speech loss.

Wetterhahn did not feel the effects of the mercury until six months after the accident. Within three months, she was dead. After her death, the U.S. Occupational Safety and Health Administration urged scientists to wear highly resistant laminate gloves (consisting of several bonded layers) under a pair of heavy-duty neoprene gloves when handling compounds such as dimethyl mercury. The thickness of the outer layer of the epidermis, combined with the toughness provided by keratin, also prevents microorganisms and viruses from entering the body.

In addition, sebum secreted by the sebaceous glands helps prevent microorganisms from living and growing on the skin surface. Since it is slightly acidic, sebum creates a condition in which many microorganisms cannot exist. Sebum serves a further protective function by keeping the skin and hair moist;

dry skin would crack, allowing viruses and bacteria to enter. If the protective outer layer of the skin is broken because of an injury and microorganisms enter the body, the many blood vessels in the dermis help prevent the microorganisms from reaching internal tissues.

As an immune response, the vessels dilate or expand. This increases the amount of blood flowing to the area, which in turn brings in more white blood cells and other protein factors to battle the infection. Even though the skin forms a protective barrier, it is still slightly permeable or allows certain substances to pass through it. Vitamins A, D, E, and K all pass through the skin and are absorbed in the capillaries in the dermis. Steroid hormones such as estrogen and chemicals such as nicotine also pass through and are absorbed. With this in mind, medical researchers have developed therapeutic patches that are attached to the skin to deliver chemicals or medication (nicotine patches for those individuals trying to quit smoking are an example).

Nails protect the exposed tips of fingers and toes from physical injury. Fingernails also aid the fingers in picking up small objects. Hair serves a protective function, although it is limited. On the head, hair protects the scalp from damaging ultraviolet (UV) radiation from the Sun, cushions the head from physical blows, and insulates the scalp to a degree. On the eyelids, eyelashes prevent airborne particles and insects from entering the eyes. Hairs in the nostrils and the external ear canals perform a similar function.

When stimulated by cold or an emotion such as fear, the arrector pili muscles contract, pulling hair follicles upright. In animals (and in our evolutionary ancestors, who had much more body hair), this action adds warmth by adding a layer of insulating air to the fur. In present-day humans, who have very little body hair, this action seems to serve no purpose other than to create dimples or 'goose bumps' in the skin.

The body is protected against the Sun's harmful UV radiation by melanin, produced by melanocytes in the epidermis. Melanin accumulates within the cells of the epidermis. It then absorbs UV radiation before that radiation can destroy the cells' DNA or deoxyribonucleic acid (large, complex molecules found in the nuclei of cells that carries genetic or hereditary information for an organism's development). Increased exposure to the Sun causes melanocytes to increase their production of melanin. The temporary result is that the skin becomes darker or tanned and is able to withstand further exposure to UV rays. The protection afforded by melanin, however, is limited. Prolonged or excessive exposure to UV radiation eventually damages the skin. It causes elastic fibres in the dermis to clump, and the skin takes on a leathery appearance. Overexposure can also result in melanoma, a tumor composed of melanocytes.

Body Temperature

Normal internal body temperature averages approximately 98.6°F (37°C).

The heat-regulating functions of the body are extremely important. If the internal temperature varies more than a few degrees from normal, life-threatening changes take place in the body. Eccrine glands play an important part in maintaining normal body temperature. When the temperature of the body rises due to physical exercise or environmental conditions, the hypothalamus (region of the brain containing many control centres for body functions and emotions) sends signals to the eccrine glands to secrete sweat. When sweat evaporates on the skin surface, it carries large amounts of body heat with it and the skin surface cools.

Because blood carries heat (a form of energy), blood flow is another regulator of body temperature. Under warm conditions, the hypothalamus signals blood vessels in the dermis to dilate or expand. This increases blood flow (and carries excess heat) to the body's surface. Like a radiator, the skin then gives off heat to the surrounding environment. During cold conditions, the hypothalamus signals eccrine glands to stop secreting sweat. It also signals blood vessels in the dermis to constrict or close, which reduces blood flow to the skin surface. As a result, heat is kept within the core of the body.

Excretion and Vitamin D Formation

Excretion is a very minor function of the skin. Sweat does contain salt and urea (a compound produced when the liver breaks down amino acids), but the amounts of these wastes are slight. The kidneys are mainly responsible for removing waste products from the blood.

SANDBLASTING YOUR FACE

For years, workers have cleaned old stone and concrete structures by blasting their surfaces with a spray of fine sand. In the late 1990s, dermatologists and beauty salon owners in the United States began using a similar technique to remove the signs of aging on people's faces. The new treatment, already used in Europe since the early 1990s, is called microdermabrasion. A machine blows tiny sterile sand crystals onto the skin of the face, then suctions them off. The crystals rub off the top layer of the skin, helping remove wrinkles.

The procedure is relatively painless and quick. However, its effects are not permanent, and it only removes fine lines. Deep lines around the mouth, crow's feet around the eyes, and deep lines on the forehead remain, although they are softened. As explained earlier, too much sunlight is harmful to the body. A limited amount, however, is beneficial. In the lower layers of the epidermis, cells contain a form of cholesterol (fatlike substance produced by the liver that is an essential part of cell membranes and body chemicals). When exposed to UV radiation, that cholesterol changes into vitamin D, which the body uses to absorb calcium and phosphorus from food in the small intestine. Those two minerals are then used to build and maintain bones and teeth, among other functions.

Sensory Reception

The main function of the sensory receptors in the dermis is to provide the brain with information about the external world and its effect on the skin. Thus, they alert the body to the possible tissue-damaging effects of extreme heat or cold or something that is pressing hard against the skin. They also transmit pleasant sensations, such as a gentle breeze blowing across the face or the soft caress of a loved one.

The receptors differ in their sensitivity. Touch receptors are the most sensitive, responding to the slightest contact. Found mainly in the fingers, tongue, and lips, they number about 500,000. Pain receptors, however, do not react unless the stimulus is strong enough. Located all over the body, pain receptors number between three and four million. Their high numbers indicate their importance to the body. Receptors send their information to the brain to be interpreted. The brain then directs the body to respond, whether to remove itself from the situation or remain. Sensation, therefore, is a function of the brain and the nervous system.

INTEGUMENTARY DISEASE SYSTEM

Unlike some other body systems, the integumentary system quickly shows when it is afflicted by an aliment or malady. Over one thousand different aliments can affect the skin. The most common skin disorders are those caused by allergies or bacterial or fungal infections. Burns and skin cancers, although less common, are more dangerous. In some cases, they can be lethal.

Acne

Acne is a skin disease marked by pimples on the face, chest, and back. The most common skin disease, acne affects an estimated 17 to 28 million people in the United States. Although it can strike people at any age, acne usually begins at puberty and worsens during adolescence. At puberty, increased levels of androgens (male hormones) cause the sebaceous glands to secrete an excessive amount of sebum into hair follicles. The excess sebum combines with dead, sticky skin cells to form a hard plug that blocks the follicle. Bacteria that normally lives on the skin then invades the blocked follicle. Weakened, the follicle bursts open, releasing the sebum, bacteria, skin cells, and white blood cells into the surrounding tissues. A pimple then forms.

Treatment for acne depends on whether the condition is mild, moderate, or severe. The goal is to reduce sebum production, remove dead skin cells, and kill skin bacteria. In very mild cases, keeping the skin clean by washing with a mild soap is recommended. In other cases, medications applied directly to the skin or taken orally may be prescribed in combination with gentle cleansing.

Athlete's Foot

Athlete's foot is a common fungus infection in which the skin between the toes becomes itchy and sore, cracking and peeling away. Properly known as tinea pedis, the infection received its common name because the infectioncausing fungi grow well in warm, damp areas such as in and around swimming pools, showers, and locker rooms (areas commonly used by athletes). The fungi that cause athlete's foot are unusual in that they live exclusively on dead body tissue (hair, the outer layer of skin, and nails). Researchers do not know exactly why some people develop the condition and others do not. It is known that sweaty feet, tight shoes, and the failure to dry feet well after swimming or bathing all contribute to the growth of the fungus.

Symptoms of athlete's foot include itchy, sore skin on the toes, with scaling, cracking, inflammation, and blisters. If the blisters break, raw patches of tissue may be exposed. If the infection spreads, itching and burning may increase. Athlete's foot usually responds well to treatment. Simple cases are treated with antifungal creams or sprays. In more severe cases, an oral antifungal medication may be prescribed.

Burns

There are few threats more serious to the skin than burns. Burns are injuries to tissues caused by intense heat, electricity, UV radiation (sunburn), or certain chemicals (such as acids). When skin is burned and cells are destroyed, the body readily loses its precious supply of fluids. Dehydration can follow, leading to a shutdown of the kidneys, a life-threatening condition. Infection of the dead tissue by bacteria and viruses occurs one to two days after skin has been burned. Infection is the leading cause of death in burn victims. Burns are classified according to their severity or depth: first-, second-, or third-degree burns. First-degree burns occur when only the epidermis is damaged. The burned area is painful, the outer skin is reddened, and slight swelling may be present. Sunburns are usually first-degree burns. Although they may cause discomfort, these minor burns are usually not serious and heal within a few days.

Second-degree burns occur when the epidermis and the upper region of the dermis are damaged. The burned area is red, painful, and may have a wet, shiny appearance because of exposed tissue. Blisters may form. These moderate burns take longer to heal. If the blisters are not broken and care is taken to prevent infection, the burned skin may regenerate or regrow without permanent scars. Third-degree burns occur when the entire depth of skin is destroyed. Because nerve endings have been destroyed, the burned area has no sensitivity. The area may be blackened or gray-white in colour. Muscle tissue and bone underneath may be damaged. In these serious to critical burns, regeneration of the skin is not possible. Skin grafting—taking a piece of skin from an unburned portion of the burn victim's body and transplanting it to the burned area—must be done to cover the exposed tissues. Third-degree burns take weeks to heal and will leave permanent scarring.

Dermatitis

Dermatitis is any inflammation of the skin. There are many types of dermatitis and most are characterized by a pink or red rash that itches. Two common types are contact dermatitis and seborrheic dermatitis. Contact dermatitis is an allergic reaction to something that irritates the skin. It usually appears within forty-eight hours after touching or brushing against a substance to which the skin is sensitive. The resin in poison ivy, poison oak, and poison sumac is the most common source of contact dermatitis. The skin of some people may also be irritated by certain flowers, herbs, and vegetables. Chemical irritants that can cause contact dermatitis include chlorine, cleaners, detergents and soaps, fabric softeners, perfumes, glues, and topical medications (those applied on the skin). Contact dermatitis can be treated with medicated creams or ointments and oral antihistamines and antibiotics.

ARTIFICIAL SKIN

Artificial skin, the synthetic or manmade equivalent of human skin, was first developed in the 1970s. Since then, the lives of many severely burned people have been saved through the use of artificial skin. In the 1970s, John F. Burke, chief of trauma services at Massachusetts General Hospital in Boston, and Ioannis V. Yannas, chemistry professor at Massachusetts Institute of Technology in Cambridge, teamed up to develop some type of human skin replacement. In their research, the two men found that collagen fibres (protein found in human skin) and a long sugar molecule (called a polymer) could be combined to form a porous material that resembles skin. They then created a kind of artificial skin using polymers from shark cartilage and collagen from cowhide.

Burke and Yannas soon discovered that artificial skin acts like a framework onto which new skin tissue and blood vessels grow. As the new skin grows, the cowhide and shark substances from the artificial skin are broken down and absorbed by the body. In 1979, Burke and Yannas used their artificial skin on their first patient, a woman who had suffered burns over half her body. After peeling away her burned skin, Burke applied a layer of artificial skin and, where possible, grafted or added on some of her own unburned skin. Three weeks later, the woman's new skin, the same colour as her unburned skin, was growing at an amazingly healthy rate.

With continued research and development, synthetic skin may become a more common treatment for burns and other serious skin disorders. Seborrheic dermatitis, known commonly as seborrhea, appears as red, inflamed skin covered by greasy or dry scales that may be white, yellow, or gray. These scaly lesions appear usually on the scalp, hairline, and face. Dandruff is a mild form of seborrheic dermatitis. Medical researchers do not know the exact cause of this skin disease. They believe that a high-fat diet, alcohol, stress, oily skin, infrequent shampooing, and weather extremes (hot or cold) may play some role. The disease may be treated with special shampoos that help soften and remove the scaly lesions. In more severe cases, medicated creams or shampoos containing coal tar may be prescribed.

Psoriasis

Psoriasis is a chronic (long-term) skin disease characterized by inflamed lesions with silvery-white scabs of dead skin. The disease affects roughly four million people in the United States, women slightly more than men. It is most common in fair-skinned people. Normal skin cells mature and replace dead skin cells every twenty-eight to thirty days. Psoriasis causes skin cells to mature in less than a week. Because the body cannot shed old skin as rapidly as new cells are rising to the surface, raised patches of dead skin develop. These patches are seen on the arms, back, chest, elbows, legs, folds between the buttocks, and scalp. The cause of psoriasis is unknown. In some cases, it may be hereditary or inherited. Attacks of psoriasis can be triggered by injury or infection, stress, hormonal changes, exposure to cold temperature, or steroids and other medications. The treatment for psoriasis depends on its severity. Steroid creams and ointments are commonly used to treat mild or moderate psoriasis. If the case is more severe, these medications may be used in conjunction with ultraviolet light B (UVB) treatments. Strong medications are reserved for those individuals suffering from extreme cases of psoriasis.

Skin Cancer

Skin cancer is the growth of abnormal skin cells capable of invading and destroying other cells. Skin cancer is the single most common type of cancer in humans. The cause of most skin cancers or carcinomas is unknown, but overexposure to ultraviolet radiation in sunlight is a risk factor. Basal cell carcinoma is the most common form of skin cancer, accounting for about 75 percent of cases. It is also the least malignant or cancerous (tending to grow and spread throughout the body). In this form of skin cancer, basal cells in the epidermis are altered so they no longer produce keratin. They also spread, invading the dermis and subcutaneous layer. Shiny, dome-shaped lesions develop most often on sunexposed areas of the face. The next most common areas affected are the ears, the backs of the hands, the shoulders, and the arms. When the lesion is removed surgically, 99 percent of patients recover fully.

Squamous cell carcinoma affects the cells of the second deepest layer of the epidermis. Like basal cell carcinoma, this type of skin cancer also involves skin exposed to the sun: face, ears, hands, and arms. The cancer presents itself as a small, scaling, raised bump on the skin with a crusting centre. It grows rapidly and spreads to adjacent lymph nodes if not removed. If the lesion is caught early and removed surgically or through radiation, the patient has a good chance of recovering completely.

Malignant melanoma accounts for about 5 percent of all skin cancers, but it is the most serious type. It is a cancer of the melanocytes, cells in the lower epidermis that produce melanin. In their early stages, melanomas resemble moles. Soon, they appear as an expanding brown to black patch. In addition to invading surrounding tissues, the cancer spreads aggressively to other parts of the body, especially the lungs and liver. Overexposure to the Sun may be a cause of melanomas, but the greatest risk factor seems to be genetic. Early discovery of the melanoma is key to survival. The primary treatment for this skin cancer is the surgical removal of the tumor or diseased area of skin. When the melanoma has spread to other parts of the body, it is generally considered incurable.

Vitiligo

Vitiligo is a skin disorder in which the loss of melanocytes (cells that produce the colour pigment melanin) results in patches of smooth, milky white skin. This often inherited disorder affects about 1 to 2 percent of the world's population. Although it is more easily observed in people with darker skin, it affects all races. It can begin at any age, but in 50 percent of the cases it starts before the age of twenty. Medical researchers do not know the exact cause of the disorder. Some theorize that nerve endings in the skin may release a chemical that destroys melanocytes. Others believe that the melanocytes simply self-destruct. Still others think that vitiligo is a type of autoimmune disease, in which the body targets and destroys its own cells and tissues. Vitiligo cannot be cured, but it can be managed. Cosmetics can be applied to blend the white areas with the surrounding normal skin. Sunscreens are useful to prevent the burning of affected areas and to prevent normal skin around the patches from becoming darker.

Warts

Warts are small growths caused by a viral infection of the skin or mucous membrane. The virus infects the surface layer. Warts are contagious. They can easily pass from person to person. They can also pass from one area of the body to another on the same person. Affecting about 7 to 10 percent of the population, warts are particularly common among children, young adults, and women. Common warts include hand warts, foot warts, and flat warts. Hand warts grow around the nails, on the fingers, and on the backs of the hands. They appear mostly in areas where the skin is broken.

Foot warts (also called plantar warts) usually appear on the ball of the foot, the heel, or the flat part of the toes. Foot warts do not stick up above the surface like hand warts. If left untreated, they can grow in size and spread into clusters of several warts. If located on a pressure point of the foot, these warts can be painful. Flat warts are smaller and smoother than other warts. They grow in great numbers and can erupt anywhere on the body. In children, they appear especially on the face.

Many nonprescription wart remedies are available that will remove simple warts from hands and fingers. Physicians use stronger chemical medications to treat warts that are larger or do not respond to over-the-counter treatments. Freezing warts with liquid nitrogen or burning them with an electric needle are advanced treatment methods.

BOVINE VIRAL DIARRHEA VIRUS

Bovine viral diarrhea virus (BVDV) is an RNA virus classified as a Pestivirus in the family Flaviviridae. The role of BVDV in BRD has been controversial, but appears to be that of a virus capable of inducing immunosuppression, which allows for the development of secondary bacterial pneumonia. Seroconversion to BVDV has been reported to be predictive of the occurrence of respiratory disease in feedlot calves, and BVDV has been reported to be the virus most frequently associated with multiple viral infections of the respiratory tract of calves.

Treatment for BVDV infection is supportive and includes antimicrobials to prevent or treat bacterial pneumonia. General principles of control are discussed under enzootic pneumonia of calves and shipping fever pneumonia. Inactivated and modified live vaccines are available for IM administration. Recently, vaccines containing both the type I and type II genotypes have become available. Modified live vaccines can induce immunosuppression and should be used with caution in highly stressed cattle. Modified live BVDV vaccines are not approved for use in pregnant cattle.

OTHER BOVINE RESPIRATORY VIRUSES

Several other viruses may potentially be involved in BRD. Bovine herpesvirus-4 has been implicated in several diseases, including BRD. Bovine adenovirus has been associated with a wide spectrum of diseases, with bovine adenovirus type 3 being the serotype most often associated with BRD. Two serotypes of bovine rhinovirus have been recognized to cause respiratory tract infections in cattle. Other viruses reported to be associated with BRD include bovine reovirus, enterovirus, and coronavirus.

There is growing evidence that bovine coronavirus may have a more important role in BRD than previously recognized. These viruses have a role similar to the other viruses previously discussed in that, in combination with other stressors, they can serve as initiators of bacterial pneumonia. Vaccines are not available for prevention of these viral respiratory diseases.

Enzootic Pneumonia of Calves and Shipping fever Pneumonia

Enzootic pneumonia and shipping fever pneumonia share many similarities in their respective etiologies and pathogeneses and general measures for control and prevention.

ENZOOTIC PNEUMONIA OF CALVES

Enzootic pneumonia of calves refers to infectious respiratory disease in calves. The term 'viral pneumonia of calves' is sometimes used but is not preferred based on the current understanding of etiology and pathogenesis. Enzootic pneumonia is primarily a problem in calves <6 mo old with peak occurrence from 2-10 week, but may be seen in calves up to 1 yr of age. It is more common in dairy than in beef calves and is a common problem in veal calves. It is also more common in housed calves than those raised outside. Peak incidence of disease may coincide with decline of passively acquired immunity. Morbidity rates may approach 100 per cent; case fatality rates vary but can reach 20 per cent.

Etiology

The etiology is similar to that for BRD complex in general. The pathogenesis involves stress and possibly an initial respiratory viral infection followed by a secondary bacterial infection of the lower respiratory tract. Stress results from environmental and management factors, including inadequate ventilation, continually adding calves to an established group, crowding, and nutritional factors such as poor-quality milk replacers. Partial or complete failure of passive transfer of maternal antibodies is an important host factor related to development of disease. Any of several viruses may be involved, and a variety of bacteria may be recovered from affected calves. Mycoplasmal and bacterial agents including Pasteurella multocida, Mannheimia haemolytica, and Mycoplasma bovis represent the most frequently isolated pathogenic organisms. The individual viral and bacterial etiologies, clinical signs, lesions, and treatment are discussed under viral respiratory tract infections and bacterial pneumonia.

Control and Prevention

When calves of varying ages are placed in communal pens, control of enzootic pneumonia is difficult. The severity of the pneumonia may be decreased by improved husbandry, proper housing, adequate ventilation, and good nursing care. Prevention begins with vaccinating the cows against specific respiratory viruses and bacteria 3-4 week prepartum to improve the quality of colostral antibodies. Calves should receive good quality colostrum at 8-10 per cent of body wt in the first 12 hr after birth. Newborn dairy calves should be housed individually in hutches or stalls and fed whole milk or a high-quality milk replacer with a fibre content of <0.25 per cent until 8-12 week old. Calves should be vaccinated against respiratory viruses 3-4 week before the first grouping, although in some situations, the presence of passive immunity may interfere with an active immune response. Calves should be of similar age when assembled into groups and the group should be limited to d'10. As calves mature, groups can become larger as the size of the herd, facilities, and available labour dictate. An 'all in/all out' management style should be practiced when establishing and terminating a group. At minimum, newly purchased calves should be isolated before introduction to an existing group. Newborn beef calves and their dams should be moved from concentrated calving areas as soon as the calf is nursing well and is strong enough to travel.

SHIPPING FEVER PNEUMONIA

Shipping fever pneumonia is a respiratory disease of cattle of multifactorial etiology with Mannheimia haemolytica and, less commonly, Pasteurella multocida or Histophilus somni, being the important infectious agents involved. Shipping fever pneumonia is associated with the assembly into feedlots of large groups of calves from diverse geographic, nutritional, and genetic backgrounds. Disease is typically seen in feeder calves 7-10 days after assembly in a feedlot. Morbidity can approach 35 per cent; mortality is 5-10 per cent.

Etiology

The pathogenesis of shipping fever pneumonia involves stress factors, with or without viral infection, interacting to suppress host defence mechanisms, which allows the proliferation of commensal bacteria in the upper respiratory tract. Subsequently, these bacteria colonize the lower respiratory tract and cause a bronchopneumonia with a cranioventral distribution in the lung. Multiple stress factors are believed to contribute to the suppression of host defence mechanisms. Transportation over long distances serves as a stressor; it may be associated with exhaustion, starvation, dehydration, chilling and overheating depending on weather conditions, and exposure to vehicle exhaust fumes. Additional stressors include passage through auction markets; commingling, processing, and surgical procedures on arrival at the feedlot; dusty environmental conditions; and nutritional stress associated with a change to high-energy rations in the feedlot. The individual viral and bacterial etiologies, clinical signs, lesions, and treatment are discussed under viral respiratory tract infections.

Control and Prevention

Prevention of shipping fever pneumonia should focus on reduction of the stressors that contribute to development of the disease. Cattle should be assembled rapidly into groups, and new animals should not be introduced to established groups. Auction markets and mixing of cattle from different sources should be avoided if possible. Transport time should be minimized, and rest periods, with access to feed and water, should be provided during prolonged transport. Calves should be weaned 2-3 week before shipment, and surgical procedures should be performed in advance of transport. Cattle should be processed within 48 hr after arrival at the feedlot. A rest period of 6-12 hr after transport may allow for rehydration and return of cortisol to levels that will have less impact on the immune response to vaccination. Adaptation to high-energy rations should be gradual as acidosis, indigestion, and anorexia may inhibit the immune response. Vitamin and mineral deficiencies should be corrected. Dust control measures should be used. Metaphylaxis with long-acting antibiotics given 'on arrival' for cattle at high risk for developing shipping fever pneumonia has been shown to significantly reduce morbidity and improve rate of gain. The administration of viral respiratory vaccines on entry to the feedlot has been historically controversial, especially with modified live vaccines. These vaccines have been reported to increase the mortality associated with shipping fever pneumonia. Continuous improvements in modified live vaccine production, and the fact that these vaccines do not require a booster, have made them preferred over killed vaccines for on-arrival processing. When possible, vaccinations for the viral and bacterial components of shipping fever pneumonia should be given 2-3 week before transport and can be repeated on entry to the feedlot.

INTERSTITIAL PNEUMONIA

This classification represents a group of respiratory diseases that are characterized by an acute onset of respiratory distress and a combination of lung lesions that include pulmonary edema and congestion, interstitial emphysema, alveolar epithelialization, and hyaline membrane formation.

Acute bovine pulmonary emphysema and edema (ABPEE) is one of the more common causes of acute respiratory distress in cattle, particularly adult beef cattle, and is characterized by sudden onset, minimal coughing, and a course that ends fatally or improves dramatically within a few days.

It is a disease involving groups of cattle; morbidity may be >50 per cent, although usually only a small minority develops severe respiratory distress. Typically, ABPEE occurs in fall, 5-10 days after change to a better, often lush, pasture. A similar condition has been reported on a wide variety of grasses, alfalfa, rape, kale, and turnip tops.

Etiology

Metabolites of the naturally occurring amino acid L-tryptophan probably are responsible for many outbreaks. In the rumen, L-tryptophan is degraded to indoleacetic acid, which can be converted to 3-methylindole by some ruminal microorganisms. 3-methylindole is absorbed into the bloodstream and is the source of the pneumotoxicity after metabolism by the mixed function oxidase system, which is very active in the lungs. Apparently, the level of L-tryptophan in crops is most likely to be high in lush, rapidly growing pastures, particularly (but not exclusively) in the fall.

Clinical Findings

ABPEE is most common in heavy beef cows but may occur in either sex and in dairy or beef cattle under similar management conditions. Nursing calves are unaffected. Outbreaks usually develop within 5-10 days of a change to better grazing and rarely occur in animals that have been on a field >3 wk. Mild cases may go unnoticed. Cattle are subdued but still alert; there is tachypnea and hyperpnea, but auscultation is usually unrewarding. Such cattle usually recover spontaneously within days.

Severely affected cattle show extensive respiratory distress with mouth breathing, extension of the tongue, and drooling. A loud expiratory grunt is common, but coughing is unusual. In the early stages, auscultation reveals surprisingly soft respiratory sounds. Mild exercise increases dyspnea and may precipitate death.

If death does not occur, the animals improve dramatically and resume eating by the third day. At this stage, auscultation reveals harsh respiratory sounds and, in some animals, dorsal (emphysematous) crackles. Some cattle have subcutaneous emphysema extending along the back from the withers. Full clinical recovery may require 3 wk.

Lesions

In affected cattle that have died or been slaughtered in extremis, the lungs are heavy and do not collapse normally. They are widely affected with various degrees of firmness; there is extensive edema and emphysema, often with the formation of large air-filled bullae in interlobular and subpleural regions. Submucosal hemorrhages are often present on the larynx and in the trachea and larger bronchi. Histologically, the lesion is characterized by congestion, alveolar edema, hyaline membrane formation, and areas of early alveolar epithelial hyperplasia of type II pneumocytes; occasionally, areas of bronchiolar necrosis may be found. The emphysema is often dramatic and is limited to interstitial fascia where it is accompanied by edema.

In animals that are slaughtered after 3 days of illness, the lungs are still heavy and do not collapse normally. They are pinkish gray and of increased firmness; edema and emphysema are inconspicuous or absent. Histologically, widespread alveolar epithelial hyperplasia characteristic of a diffuse, acute, proliferative alveolitis is seen.

Diagnosis

Diagnosis is based on history, signs, and lesions. Because the syndrome is not specific with regard to cause, evidence must be obtained from management factors such as change in pasture.

Treatment

Severely affected animals have so little pulmonary reserve that any driving or handling must be done with caution to prevent immediate deaths. Removal of cattle from the offending pastures may not prevent the development of new cases for the next 4-7 days. No treatment has been identified that will reverse the fully developed lesions of ABPEE.

Control

One approach to control is dietary management, including the following options: 1) avoiding pastures likely to induce ABPEE, 2) feeding hay before turn out on pasture and limiting exposure time on suspect pastures, 3) limiting grazing time and gradually increasing exposure to the pasture over time, 4) using pastures before they become lush, 5) delaying use of lush pastures until after a hard frost, 6) initially grazing pastures with less susceptible stock (cattle <15 mo of age or sheep), or 7) using strip grazing.

A medical approach to control involves feeding monensin or lasalocid, which inhibit the bacteria that convert L-tryptophan to 3-methylindole. Treatment with monensin can be started 1 day before introduction to pasture, whereas lasalocid requires a 6-day pretreatment period. These drugs are of no benefit after onset of clinical signs.

ANAPHYLAXIS

Anaphylaxis or Type I hypersensitivity reactions in cattle can result in an atypical interstitial pneumonia. The lung is a major target organ in cattle for Type I hypersensitivity. Clinical signs are those of acute respiratory distress. Cattle that die of anaphylaxis may have lesions consistent with those described for atypical interstitial pneumonia.

Treatment is the administration of epinephrine; supportive treatment includes anti-inflammatory therapy with corticosteroids or NSAID. If pharyngeal or laryngeal edema is present, a tracheostomy may be indicated.

HYPERSENSITIVITY PNEUMONITIS

A condition that appears to be similar to farmer's lung disease in humans occurs in both acute and chronic forms in adult cattle. The human and bovine forms of the disease may coexist on problem farms due to common exposure to dust from moldy hay.

Etiology

The disease occurs when sensitized individuals inhale antigens from thermophilic actinomycetes, commonly the spores of Micropolyspora faeni. The actinomycetes proliferate in vast numbers in hay, grain, or other vegetable material that has overheated to ~150°F (65°C) after damp storage (30-40 per cent moisture content). Dust that contains large numbers of spores is released when this moldy hay is shaken. The small size (1 μm) of the spores allows them to reach the smallest airways and alveoli to provoke a reaction that has been termed a 'hypersensitivity pneumonitis'; this is considered to be predominantly a Type III hypersensitivity reaction, although a Type IV hypersensitivity component is suspected.

Affected herds exist in areas where significant rainfall usually occurs during the haymaking season, suggesting that a clinical problem may arise only after repeated sensitization and challenge from the spores. Clinical disease tends to arise during the latter half of the winter feeding period and usually only when moldy hay is fed indoors. Under such circumstances, serum antibodies (usually detected by immunodiffusion) to M faeni are widespread among adult cattle by the end of each winter feeding period, and many apparently normal cattle are seropositive. By contrast, few adult cattle are seropositive on other farms on which 'good' hay or grass silage is fed.

Clinical Findings

Cattle may succumb to the acute form of the disease over a period of weeks. Usually, only severe acute cases are noticed. There is respiratory distress, anorexia, and agalactia in animals e'5 yr old; coughing and pyrexia also occur, and adventitious sounds are occasionally heard on auscultation. Death is rare. The chronic disease usually has a higher morbidity; in most instances, the signs are weight loss, poor production, and persistent coughing. Affected cattle are fairly bright and eat reasonably well, but tachypnea, hyperpnea, and coughing are widespread. Auscultation may reveal cranioventral crackles and sometimes, in more severe cases, scattered rhonchi. Exercise intolerance may be seen, and congestive cardiac failure can develop if pulmonary fibrosis is widespread.

Lesions

The macroscopic lesions are often unremarkable; usually, there is mild peripheral lobular overinflation with diffusely scattered, small, gray, subpleural spots. Although transient pulmonary edema may be a feature of severe acute cases, the histologic lesions that are consistently found are interalveolar cellular infiltration, epithelioid granulomata, and bronchiolitis obliterans. In some chronic cases, small foci of alveolar epithelial hyperplasia and metaplasia with interstitial fibrosis are found. These areas may extend to include most, if not all, of the lung substance to produce cases clinically indistinguishable from diffuse fibrosing alveolitis. Circumstantial evidence suggests that some cases of diffuse fibrosing alveolitis are the end stage of hypersensitivity pneumonitis.

Treatment and Control

Because it is often impossible to completely shield cattle from further challenge, most recover only partially after dexamethasone treatment (1 mg/ 5-10 kg body wt). However, improvement is usually marked when cattle are turned out in the spring. Prevention is difficult in areas where hay is likely to be wet during the curing process and it is not possible to alter the feeding regimen.

DIFFUSE FIBROSING ALVEOLITIS

Diffuse fibrosing alveolitis is a chronic, progressive respiratory disease of undetermined cause and possibly of multiple etiologies. A proportion of affected cattle are seropositive for precipitating antibodies to Micropolyspora faeni, and this condition may represent the end stage of hypersensitivity pneumonitis. Other than the respiratory signs, the animals appear alert and maintain a good appetite until the onset of heart failure in the terminal stages. Signs include coughing, increased respiratory rate, dyspnea, and weight loss. Necropsy findings include right ventricular hypertrophy, interalveolar fibrosis, obliteration of the alveolar spaces, alveolar hyperplasia, bronchitis, and bronchiolitis. There is no treatment.

ACUTE RESPIRATORY DISTRESS SYNDROME OF FEEDLOT CATTLE

An acute respiratory distress syndrome has been described in feedlot cattle with clinical signs and pathologic findings of an atypical interstitial pneumonia. The syndrome occurs sporadically and the etiology remains undefined. Bovine respiratory syncytial virus, abnormal production of 3-methylindole in the rumen, dusty conditions, and pre-existing lesions of chronic cranioventral bacterial pneumonia have been suggested as causes or contributing factors. Clinical signs include respiratory distress characterized by tachypnea and dyspnea, and affected cattle may be found dead if clinical signs are unobserved. Lesions are those of atypical interstitial pneumonia with prominent emphysema and edema in the lungs. Treatment protocols have not been defined, and thus would be symptomatic and supportive. Management strategies suggested include vaccinating for bovine respiratory syncytial virus, controlling dust in the feedlot, and avoiding abrupt dietary changes.

4-Ipomeanol Toxicity (Moldy Sweet Potato) and Perilla Ketone Toxicity (Purple Mint Toxicity)

Clinicopathologic syndromes indistinguishable from acute bovine pulmonary emphysema and edema occur after ingestion of either moldy sweet potatoes infested with Fusarium solani, or the wild mint Perilla frutescens. Moldy sweet potato toxicity is caused by the ingestion of a furanoterpenoid toxin produced by sweet potatoes (Ipomoea batatus) in response to infestation with the fungus F solani; the end result is production of the pneumotoxin 4-ipomeanol. Perilla ketone toxicity is caused by ingestion of the leaves and seeds of the plant P frutescens (purple mint), which contains a pneumotoxin and is found in the southeastern USA. The pathogeneses of both these conditions are similar to that of ABPEE, as is approach to treatment.

BODY ART AND MUTILATION IN DISEASE

Tattoos are relatively permanent marks or designs made on the skin. Tattoo comes from the Tahitian word *tattau*, meaning 'to mark.' The process of tattooing is accomplished by injecting coloured pigment into small deep holes made in the skin. The modern method of tattooing employs an electric needle to inject the pigment. People have been decorating their bodies with pictures of animals, flowers, supernatural creatures, and various designs for thousands of years. Egyptian mummies dating from 3035 b.c. have been discovered with ornate designs of flowers tattooed on their skin. Many ancient cultures believed that a tattoo of an animal could capture the mystical spirit of that animal and magically link the wearer to the animal depicted. While many cultures have revered tattoos, many others have considered them vulgar and offensive. For as long as people have applied tattoos to their skin, they have sought ways to remove them. In modern times, tattoos can be removed medically through one of four ways. If the tattoo is small, it can be surgically cut off and the skin sewn back together. In a method called dermabrasion, the tattoo is 'sanded' with a rotary abrasive instrument until the layers of skin peel. Another method that uses abrasion is called salabrasion. In this procedure, which is centuries old, salt water is applied to the tattoo and then it is vagourously rubbed with some sort of sanding device until the tattoo pigments are dispersed. All three of these methods leave some sort of scarring, but the last method, laser surgery, does not. Pulses of light from a laser are directed onto the tattoo, breaking up its pigments. The pigments are then removed over the next few weeks by the body's defence cells.

KEEPING THE INTEGUMENTARY SYSTEM HEALTHY

The epidermis thins as basal cells divide less and less. The dermis also thins and its elastic fibres decrease in size. As a result, the skin becomes weaker and starts to sag, forming wrinkles. Melanocytes decrease production of melanin, and the skin becomes pale and hair turns white. Sebaceous glands also decrease production of sebum, causing the skin to become dry and scaly. Blood supply to the skin is reduced and body temperature cannot be regulated as well. Finally, the skin takes longer and longer to repair itself. Although there is no way to avoid aging of the skin, there are ways to decrease the effects of aging. The loss of elasticity in the skin is speeded up by sunlight. The skin should be shielded from the Sun through the use of sunscreens, sunblocks, and protective clothing. Sunburns are never healthy and should always be avoided. This will also help reduce the risk of skin cancer.

As in all other body systems, the following play a part in keeping the integumentary system operating at peak efficiency: proper nutrition, healthy amounts of good-quality drinking water, adequate rest, regular exercise, and stress reduction. Hair loss and graying are both genetically controlled, but stress can add to both conditions. Exercise and relaxation techniques are proven ways to reduce stress. Proper daily cleansing of the skin is highly recommended. However, harsh detergents and scrubbing will not make the skin cleaner. In fact, they can injure the skin and cause excessive drying. Greater benefits can be gained by cleaning the skin with gentle soaps or lotions, then applying an appropriate moisturizer to all areas of the body.

RESPIRATORY DISEASES

Respiratory disease is among the most economically important diseases of cattle in production on a worldwide basis. Allergic rhinitis is an uncommon disease of cattle that, when chronic, may lead to granuloma formation. The etiology is an allergic reaction to pollen or fungal spores. Signs are seasonal and occur under warm, moist conditions; they include rhinorrhea, sneezing, and a sudden onset of dyspnea. In the chronic stage, multiple granulomas may form on the mucosal surface of the nasal cavity. Cytologic examination of nasal discharges may reveal eosinophils. Treatment should focus on removing the allergen or removing the animal from the allergen. Treatment with corticosteroids to block the hypersensitivity reaction is a consideration.

SINUSITIS

Etiology

Sinusitis in cattle typically involves the frontal or maxillary sinus. Frontal sinusitis is usually associated with dehorning and maxillary sinusitis with infected teeth. Numerous bacteria have been isolated from sinusitis infections in cattle.

Clinical Findings

Frontal sinusitis may occur immediately after dehorning while the site is still open or months later after the dehorning site has healed. The condition is most often unilateral. Signs may include anorexia, pyrexia, unilateral or bilateral nasal discharge, changes in air flow through the nasal passages, and foul breath. Head carriage may be abnormal. In longstanding cases of frontal sinusitis, there may be distortion of the frontal bone, exophthalmos, and neurologic signs.

Diagnosis

Diagnosis can usually be made on the basis of clinical signs. Percussion may reveal a dull sound over the affected sinus. Radiographs may reveal fluid in the sinus, the presence of dental disease, or bone lysis. Cytology of aspirated material from the affected sinus may reveal purulent material.

Treatment

Sinusitis is treated by draining the affected sinus. Trephine sites should be reviewed for appropriate anatomic landmarks. If an infected tooth is the cause of maxillary sinusitis, the tooth can be repelled through a sinusotomy site created with a trephine. Once drainage has been established, the sinus can be lavaged daily with antiseptic solutions. Treatment with parenteral antibiotics is indicated if systemic signs are present. NSAID can be given for pain relief, if needed. The prognosis is guarded.

Control

The best control method is to dehorn calves at a young age using a closed dehorning technique. If this is not possible, close attention should be paid to disinfection of surgical instruments between animals, dust control, and fly control.

TRACHEAL EDEMA SYNDROME OF FEEDER CATTLE

Tracheal edema syndrome is characterized by extensive edema of the mucosa and submucosa in the dorsal membrane of the lower trachea. The etiology is unknown. Proposed causes include respiratory viruses and bacteria, trauma to the trachea from feed bunks, passive congestion and edema from excessive fat accumulation in the thoracic inlet, hypersensitivity reactions, and mycotoxins. The condition occurs in heavy feeder cattle in the later two-thirds of the feeding period throughout North America but may be most severe in the summer in southern plains (USA) feedlots. Onset is sudden and appears to be associated with an increase in respirations stimulated by hot weather or exercise. The initial signs are a loud inspiratory noise (stridor) and the onset of dyspnea. Forced movement causes the respiratory distress to worsen. The cattle become cyanotic and typically collapse and die of asphyxiation in <24 hr. Usually, only 1 or 2 animals per pen are affected.

In the acute form, necropsy lesions include edematous and/or hemorrhagic thickening of the submucosa and mucosa of the dorsal trachea extending from the midcervical area to the thoracic inlet. There is extensive hemorrhage in the trachea but no lung lesions. In the chronic form, lesions consist of hyperemia of the caudal third of the trachea with mucopurulent exudate in the trachea. In fatal cases, the lesion becomes completely obstructive. Movement and handling of affected cattle should be limited. Antibiotics and corti-costeroids are recommended for the acute form. Tracheostomy may be required in severe cases. Providing shade and cooling with fans or water sprays is recommended. Animals that recover are prone to relapse and should be sent to slaughter.

BOVINE RESPIRATORY DISEASE COMPLEX

Bovine respiratory disease (BRD) has a multifactorial etiology and develops as a result of complex interactions between environmental factors, host factors, and pathogens. Environmental factors (*e.g.*, weaning, transport, commingling, crowding, and inadequate ventilation) serve as stressors that adversely affect the immune and nonimmune defence mechanisms of the host.

In addition, certain environmental factors (*e.g.,* crowding and inadequate ventilation) can enhance the transmission of infectious agents among animals. Many infectious agents have been associated with BRD. An initial pathogen (*e.g.,* a virus) may alter the animal's defence mechanisms, allowing colonization of the lower respiratory tract by bacteria.

BACTERIAL PNEUMONIA

Etiology

Mannheimia haemolytica, serotype 1 is the bacterium most frequently isolated from the lungs of cattle with BRD. Although less frequently cultured, Pasteurella multocida is also an important cause of bacterial pneumonia.

Histophilus somni is being increasingly recognized as an important pathogen in BRD; these bacteria are normal inhabitants of the nasopharynx of cattle. When pulmonary abscessation occurs, generally in association with chronic pneumonia, Arcanobacterium pyogenes is frequently isolated.

Under normal conditions, M haemolytica remains confined to the upper respiratory tract, in particular the tonsillar crypts, and is difficult to culture from healthy cattle. After stress or viral infection, the replication rate of M haemolytica in the upper respiratory tract increases rapidly, as does the likelihood of culturing the bacterium. The increased bacterial growth rate in the upper respiratory tract followed by inhalation and colonization of the lungs may occur due to suppression of the host's defence mechanism related to environmental stressors or viral infections. It is during this log phase of growth of the organism in the lungs that virulence factors are elaborated by M haemolytica, such as an exotoxin that has been referred to as leukotoxin.

The interaction between the virulence factors of the bacteria and host defences results in tissue damage with characteristic necrosis, thrombosis, and exudation and the development of pneumonia.

The pathogenesis of pneumonia caused by P multocida is poorly understood. This organism may opportunistically colonize lungs with chronically damaged respiratory defences, such as occurs with enzootic calf pneumonia or existing lung lesions of feedlot cattle, and cause a purulent bron-chopneumonia.

H somni may invade the lung and cause pneumonia following damage to the respiratory defences. This organism is capable of systemic spread from the lung to the brain, myocardium, synovium, and pleural and pericardial surfaces; often death can occur later in the feeding period from involvement of these additional organ systems.

Clinical Findings

Clinical signs of bacterial pneumonia are often preceded by signs of viral infection of the respiratory tract. With the onset of bacterial pneumonia, clinical signs increase in severity and are characterized by depression and toxemia. Fever (104-106°F [40-41°C]); serous to mucopurulent nasal discharge; moist cough; and a rapid, shallow respiratory rate may be noted. Auscultation of the cranioventral lung field reveals increased bronchial sounds, crackles, and wheezes. In severe cases, pleurisy may develop, characterized by an irregular breathing pattern and grunting on expiration. The animal will become unthrifty in appearance if the pneumonia becomes chronic, which is usually associated with the formation of pulmonary abscesses.

Lesions

M haemolytica causes a severe, acute, hemorrhagic fibrinonecrotic pneumonia. The pneumonia has a bronchopneumonic pattern. Grossly, there are extensive reddish black to grayish brown cranioventral regions of consolidation with gelatinous thickening of interlobular septa and fibrinous pleuritis. There are extensive thromboses, foci of lung necrosis, and limited evidence of bronchitis and bronchiolitis. P multocida is associated with a less fulminating fibrinous to fibrinopurulent bronchopneumonia. In contrast to M haemolytica, P multocida is associated with only small amounts of fibrin exudation, some thromboses, limited lung necrosis, and suppurative bronchitis and bronchiolitis.

H somnus infection of the lungs results in purulent bronchopneumonia that may be followed by septicemia and infection of multiple organs. Occasionally, H somni is associated with extensive pleuritis. Pulmonary abscessation can occur as the pneumonia becomes chronic. Abscesses develop in ~3 week but do not become encapsulated until 4 week. Arcanobacterium pyogenes is frequently cultured from these abscesses.

Diagnosis

Generally, neither serologic testing nor direct bacterial detection are performed, and diagnosis relies on bacterial culture. Because the bacteria involved are normal inhabitants of the upper respiratory tract, the specificity of culture can be increased by collecting antemortem specimens from the lower respiratory tract by tracheal swab, transtracheal wash, or bronchoalveolar lavage. Lung specimens can be collected for culture at postmortem. If possible, specimens for culture should be collected from animals that have not been treated with antibiotics to permit determination of antimicrobial sensitivity patterns.

Treatment

Early recognition by trained personnel skilled at detecting the early symptoms of disease and treatment with antibiotics are essential for successful therapy. Antibiotics effective against the 3 gram-negative bacteria most often involved in BRD should be selected. Responses to treatment should be monitored and periodic culture and sensitivity should be performed to aid in the selection of antibiotics. Long-acting antibiotics have been specifically developed for treating bacterial pneumonia in cattle. It is important that antibiotic therapy extend beyond apparent recovery to avoid relapses. Mass medication in feed or water is of limited value because sick animals do not eat or drink enough to achieve inhibitory blood levels of the antibiotic, and many of these oral antibiotics are poorly absorbed in ruminants. NSAID have been shown to be a beneficial ancillary therapy in treating bacterial pneumonia. If pulmonary abscessation has occurred, it is difficult to achieve resolution with antimicrobials and culling of the animal should be considered.

Control

General principles of control are discussed under enzootic pneumonia of calves and shipping fever pneumonia. The value of M haemolytica and P multocida bacterins is questionable, and some reports indicate they may even exacerbate the disease. Newer vaccines, which include live culture and subunit vaccines (leukotoxin), show much more promise for disease prevention. Vaccination should be done 3 week before transport to the feedlot and can be repeated on arrival. In dairy calves, vaccination of the dam may be of benefit by providing passive immunity to the calf. H somni bacterins are available, and there is some evidence that they are effective in control of BRD.

MYCOPLASMAL PNEUMONIA

The exact role of mycoplasmas and ureaplasmas in BRD requires better definition. Mycoplasmas can be recovered from the respiratory tract of nonpneumonic calves, but the frequency of isolation is greater in those with respiratory tract disease. Mycoplasmas commonly recovered from the lungs of pneumonic calves include Mycoplasma dispar, M bovis, and Ureaplasma spp. M bovis has been associated with otitis media in young calves and polyarthritis in feedlot cattle. Experimental infections usually result in inapparent to mild signs of respiratory disease. This does not preclude a synergistic role for mycoplasmas in conjunction with viruses and bacteria in BRD. Lesions include focal pulmonary abscessation and necrosis with histologic lesions of peribronchial and peribronchiolar lymphoid cuffing and alveolitis. Culture of these organisms requires special media and conditions; growth of the organisms may take up to a week. Mycoplasmas are sensitive to several antibiotics, including the tetracyclines and macrolides.

CHLAMYDIAL PNEUMONIA

Chlamydial agents have been implicated in a number of diseases of cattle, including pneumonia. Only mild clinical signs and lesions of bronchopneumonia have been produced by experimental infections. A synergism between Chlamydia and Mannheimia haemolytica has been demonstrated experimentally. Because this pathogen is infrequently tested for, its overall importance remains undetermined. The organism can be tested for by staining sections of lung lesions with Gimenez stain or by fluorescent antibody. Isolation requires inoculation of yolk sacs of embryonating chicks. Chlamydial agents are sensitive to tetracyclines.

CONTAGIOUS BOVINE PLEUROPNEUMONIA

This highly contagious pneumonia is generally accompanied by pleurisy. It is present in Africa, the Iberian peninsula, and parts of India and China; minor outbreaks occur in the Middle East. The USA has been free of the disease since 1892, the UK since 1898, and Australia since 1973.

Etiology

The causal organism is Mycoplasma mycoides *mycoides* small colony type. Susceptible cattle become infected by inhaling droplets disseminated by coughing in affected cattle. Goats and sheep are not important in the epidemiology. Septicemia produces lesions in the kidneys and placenta, which can be sources of infection. Transplacental infection of the fetus can occur. Viability of the organism in the environment is poor. The incubation period varies, but most cases occur 3-8 week after exposure. In some localities, susceptible herds may show up to 100 per cent morbidity, but much lower infection rates (~10 per cent) associated with clinical signs are more common. Mortality is likely to be ~50 per cent. Of recovered animals, 25 per cent may become carriers with chronic lung lesions in the form of sequestra of variable size. Because carriers may not be detectable clinically or serologically, they constitute a serious problem in control programmes. Breed susceptibility, management systems, and general health of the animal are important factors that influence the infection.

Clinical Findings

In acute cases, signs include fever up to 107°F (41.5°C), anorexia, and painful, difficult breathing. In hot climates, the animal often stands by itself in the shade, its head lowered and extended, its back slightly arched, and its elbows turned out. Percussion of the chest is painful; respiration is rapid, shallow, and abdominal. If the animal is forced to move quickly, the breathing becomes more distressed and a soft, moist cough may result. The disease progresses rapidly, animals lose condition, and breathing becomes very laboured, with a grunt at expiration. The animal becomes recumbent and dies after 1-3 wk. Chronically affected cattle usually exhibit signs of varying intensity for 3-4 week, after which the lesions gradually resolve and the animals appear to recover. Subclinical cases occur and may be important as carriers.

Lesions

The thoracic cavity may contain up to 10 L of clear yellow or turbid fluid mixed with fibrin flakes, and the organs in the thorax are often covered by thick deposits of fibrin. Varying amounts of one or both lungs may be involved, the affected portion being enlarged and solid. On section of the lung, the typical marbled appearance of pleuropneumonia is evident due to the widened interlobular septa and subpleural tissue that encloses gray, yellow, or red consolidated lung lobules. Microscopically, this is a severe, acute, fibrinous pneumonia with fibrinous pleurisy, thrombosis of pulmonary blood vessels, and areas of necrosis of lung tissue; the interstitial tissue is markedly thickened by edema fluid containing much fibrin. In chronic cases, the lesion has a necrotic centre sequestered in a thick, fibrous capsule, and there may be fibrous pleural adhesions. Organisms may survive in these sequestra, and the animals become carriers.

Diagnosis

Diagnosis is based on clinical signs, complement fixation test, and necropsy. Confirmation is by histopathology, detection of organisms in pleural fluid using darkfield microscopy, isolation of the organism from lung or pleural fluid, or demonstration of specific antigens in lung tissue by immunodiffusion or immunofluorescence and hyperimmune antigalactan serum. Subclinical disease is detected by complement fixation test. As soon as an outbreak is suspected, slaughter and necropsy of presumptively infected cattle is advisable.

Control

The disease is reportable by law in many countries from which it has been eradicated by slaughter of all infected and exposed animals. In countries where cattle movement can readily be restricted, the disease can be eradicated by quarantine, blood testing, and immunization with attenuated vaccine (*e.g.*, T1/44 strain). Where cattle cannot be confined, the spread of infection can be limited by vaccination. Tracing the source of infected cattle detected at abattoirs, blood testing, and imposition of strict rules for cattle movement also can aid in control of the disease in such areas. Treatment is recommended only in endemic areas because the organisms may not be eliminated, and carriers may develop. Tylosin (10 mg/kg, IM, bid for 6 injections) is reported to be effective.

VIRAL RESPIRATORY TRACT INFECTIONS

Etiology

Parainfluenza-3 virus (PI-3) is an RNA virus classified in the paramyxovirus family. Infections caused by PI-3 are common in cattle.

Although PI-3 is capable of causing disease, it is usually associated with mild to subclinical infections. The most important role of PI-3 is to serve as an initiator that can lead to the development of secondary bacterial pneumonia.

Clinical Findings and Lesions

Clinical signs include pyrexia, cough, serous nasal and lacrimal discharge, increased respiratory rate, and increased breath sounds.

The severity of signs worsens with the onset of bacterial pneumonia. Fatalities from uncomplicated PI-3 pneumonia are rare.

Lesions include cranioventral lung consolidation, bronchiolitis, and alveolitis with marked congestion and hemorrhage. Inclusion bodies may be identified. Most fatal cases have a concurrent bacterial bronchopneumonia.

Diagnosis

- Diagnostic procedures for PI-3 are similar to those for bovine respiratory syncytial virus.

Treatment and Prevention

Treatment focuses on the antimicrobial therapy directed towards bacterial pneumonia. NSAID are also a therapeutic consideration.

PI-3 vaccines are available and are almost always combined with bovine herpesvirus 1 (infectious bovine rhinotracheitis). Modified live and inactivated vaccines are available for IM administration. Vaccines containing temperature-sensitive mutants for intranasal administration are also available.

BOVINE RESPIRATORY SYNCYTIAL VIRUS

Etiology

Bovine respiratory syncytial virus (BRSV) is an RNA virus classified as a pneumovirus in the paramyxovirus family. This virus was named for its characteristic cytopathic effect—the formation of syncytial cells. In additional to cattle, sheep and goats can also be infected by respiratory syncytial viruses. Human respiratory syncytial virus (HRSV) is an important respiratory pathogen in infants and young children. Antigenic subtypes are known to exist for HRSV, and preliminary evidence suggests that there may be antigenic subtypes of

BRSV. BRSV is distributed worldwide, and the virus is indigenous in the cattle population.

BRSV infections associated with respiratory disease occur predominantly in young beef and dairy cattle. Passively derived immunity does not appear to prevent BRSV infections but will reduce the severity of disease. Initial exposures to the virus are associated with severe respiratory disease; subsequent exposures result in mild to subclinical disease. BRSV is an important virus in the bovine respiratory disease complex because of its frequency of occurrence, predilection for the lower respiratory tract, and ability to predispose the respiratory tract to secondary bacterial infection. In outbreaks, morbidity tends to be high, and the case fatality rate can be 0-20 per cent.

Clinical Findings and Lesions

Fever (104-108°F [40-42°C]), depression, decreased feed intake, increased respiratory rate, cough, and nasal and lacrimal discharge are common. Dyspnea, possibly with open-mouthed breathing, may become pronounced in the later stages of the disease. Subcutaneous emphysema may occur. Secondary bacterial pneumonia is a frequent occurrence. A biphasic disease pattern has been described but is not consistent. Gross lesions include a diffuse interstitial pneumonia with subpleural and interstitial emphysema along with interstitial edema. These lesions are similar to and must be differentiated from other causes of interstitial pneumonia. Bronchopneumonia of bacterial origin is usually present. Histologic examination reveals syncytial cells in bronchiolar epithelium and lung parenchyma, intracytoplasmic inclusion bodies, proliferation and/or degeneration of bronchiolar epithelium, alveolar epithelialization, edema, and hyaline membrane formation.

Diagnosis

A diagnosis of BRSV requires laboratory confirmation. BRSV is a difficult virus to detect, although chances of isolation may improve when sampling animals that are in the incubation or acute phases of infection. An antigen detection enzyme immunoassay is useful in detecting BRSV antigen and establishing a diagnosis. Other procedures that have proved useful in detection of BRSV antigen are fluorescent antibody and immunoperoxidase staining.

Paired serum samples can be used to establish a diagnosis. However, the antibody titer of animals with well-developed clinical disease may be higher in the acute sample than in the sample taken 2-3 week later because the antibody response often develops rapidly, and clinical signs follow virus infection by up to 7-10 days. Single serum samples with high antibody titers from a number of animals in a respiratory outbreak may be useful in making a diagnosis if coupled with clinical signs. Calves that become infected with BRSV in the presence of passively derived antibody may not seroconvert.

Treatment and Prevention

Treatment focuses on antimicrobial therapy to control secondary bacterial pneumonia. There is no specific treatment for the viral interstitial pneumonia. Supportive therapy and correction of dehydration may be necessary. There are anecdotal reports of treatment with antihistamines and/or corticosteroids being of benefit. Most cases will recover in several days without treatment. General control and prevention are discussed under enzootic pneumonia of calves and shipping fever pneumonia. Inactivated and modified live vaccines are available and may serve to reduce losses associated with BRSV.

BOVINE HERPESVIRUS 1

Etiology and Epidemiology

Bovine herpesvirus 1 (BHV-1) is associated with several diseases in cattle: infectious bovine rhinotracheitis (IBR), infectious pustular vulvovaginitis (IPV), balanoposthitis, conjunctivitis, abortion, encephalomyelitis, and mastitis. Only a single serotype of BHV-1 is recognized; however, three subtypes of BHV-1 have been described on the basis of endonuclease cleavage patterns of viral DNA—BHV-1.1 (respiratory subtype), BHV-1.2 (genital subtype), and BHV-1.3 (encephalitic subtype). BHV-1.3 has been reclassified as a distinct herpesvirus designated BHV-5.

BHV-1 infections are widespread in the cattle population. In feedlot cattle, the respiratory form is most common. The viral infection alone is not life-threatening but predisposes to secondary bacterial pneumonia, which may result in death. In breeding cattle, abortion or genital infections are more common. Genital infections can occur in bulls (infectious pustular balanoposthitis) and cows (IPV) within 1-3 days of mating or close contact with an infected animal. Transmission can occur in the absence of visible lesions and through artificial insemination with semen from subclinically infected bulls. Cattle with latent BHV-1 infections generally show no clinical signs when the virus is reactivated, but they serve as a source of infection for other susceptible animals.

Clinical Findings

The incubation period for the respiratory and genital forms is 2-6 days. In the respiratory form, clinical signs range from mild to severe, depending on the presence of secondary bacterial pneumonia. Clinical signs include high fever, anorexia, coughing, excessive salivation, nasal discharge that progresses from serous to mucopurulent, conjunctivitis with lacrimal discharge, inflamed nares (hence the common name 'red nose'), and dyspnea if the larynx becomes occluded with purulent material. Nasal lesions consist of numerous clusters of grayish necrotic foci on the mucous membrane of the septal mucosa, just visible inside the external nares. They may later be accompanied by pseudodiphtheritic yellowish plaques. Conjunctivitis with corneal opacity may occur as the only manifestation of BHV-1 infection. In the absence of bacterial pneumonia, recovery generally occurs 4-5 days after the onset of signs.

Abortions may occur concurrently with respiratory disease but may be seen up to 100 days after infection. They can occur regardless of the severity of disease in the dam. Abortions generally occur during the second half of pregnancy, but early embryonic death is possible.

In genital infections, the first signs are frequent urination, elevation of the tailhead, and a mild vaginal discharge. The vulva is swollen, and small papules, then erosions and ulcers, are present on the mucosal surface. If secondary bacterial infections do not occur, animals recover in 10-14 days. With bacterial infection, there may be inflammation of the uterus and transient infertility, with purulent vaginal discharge for several weeks. In bulls, similar lesions occur on the penis and prepuce.

BHV-1 infection can be severe in young calves and cause a generalized disease. Pyrexia, ocular and nasal discharges, respiratory distress, diarrhea, incoordination, and eventually convulsions and death may occur in a short period after generalized viral infection.

Lesions

In uncomplicated IBR infections, most lesions are restricted to the upper respiratory tract and trachea. Petechial to ecchymotic hemorrhages may be found in the mucous membranes of the nasal cavity and the paranasal sinuses. Focal areas of necrosis develop in the nose, pharynx, larynx, and trachea. The lesions may coalesce to form plaques.

The sinuses are often filled with a serous or serofibrinous exudate. As the disease progresses, the pharynx becomes covered with a serofibrinous exudate, and blood-tinged fluid may be found in the trachea. The pharyngeal and pulmonary lymph nodes may be acutely swollen and hemorrhagic. The tracheitis may extend into the bronchi and bronchioles; when this occurs, epithelium is sloughed in the airways. The viral lesions are often masked by secondary bacterial infections. In young animals with generalized BHV-1 infection, erosions and ulcers overlaid with debris may be found in the nose, esophagus, and forestomachs. In addition, white foci may be found in the liver, kidney, spleen, and lymph nodes. Aborted fetuses may have pale, focal, necrotic lesions in all tissues, which are especially visible in the liver.

Diagnosis

Uncomplicated BHV-1 infections can be diagnosed based on the characteristic signs and lesions. However, because the severity of disease can vary, it is best to differentiate BHV-1 from other viral infections by viral isolation. Samples should be taken early in the disease, and a diagnosis should be possible in 2-3 days. A rise in serum antibody titer also can be used to confirm a diagnosis. It is not possible to detect a rising antibody titer in abortions, because infection generally occurs a considerable length of time before the abortion, and titers are already maximal. BHV-1 abortion can be diagnosed by identifying characteristic lesions and demonstrating the virus in fetal tissues by virus isolation, immunoperoxidase, or fluorescent antibody staining. Gross and microscopic lesions detected shortly after death may help to establish a diagnosis.

Treatment and Control

Antimicrobial therapy is indicated to prevent or treat secondary bacterial pneumonia. General recommendations for control are discussed under shipping fever pneumonia. Immunization with modified live or inactivated virus vaccines generally provides adequate protection against clinical disease. Both IM and intranasal modified live vaccines are available, but the IM types may cause abortion in pregnant cattle. The intranasal vaccines can be used in pregnant cattle. The IM vaccines are easier to use and often are the vaccines of choice in feedlots. Breeding and replacement heifers and bulls should be immunized when 6-8 mo old, before breeding, and yearly thereafter. Some recommend that young bulls not be vaccinated because they may be discriminated against when sold for breeding if they have antibody titers. Feeder calves should be immunized 2-3 week before entry into the feedlot. Eradication of the virus is possible by serologic testing and either culling reactors or running a strict 2-herd system. To aid in eradication, deletion mutant vaccines have been developed that permit discrimination between antibody produced in response to the vaccine and antibody produced in response to natural exposure.

2

Bacterial Associations with Animals

Bacteria are consistently associated with the body surfaces of animals. There are many more bacterial cells on the surface of a human (including the gastrointestinal tract) than there are human cells that make up the animal. The bacteria and other microbes that are consistently associated with an animal are called the normal flora, or more properly the "indigenous microbiota", of the animal. These bacteria have a full range of symbiotic interactions with their animal hosts.

In biology, symbiosis is defined as "life together", *i.e.*, that two organisms live in an association with one another. Thus, there are at least three types of relationships based on the quality of the relationship for each member of the symbiotic association.

TYPES OF SYMBIOTIC ASSOCIATIONS

- *Mutualism:* Both members of the association benefit. For humans, one classic mutualistic association is that of the the lactic acid bacteria that live on the vaginal epithelium of a woman.

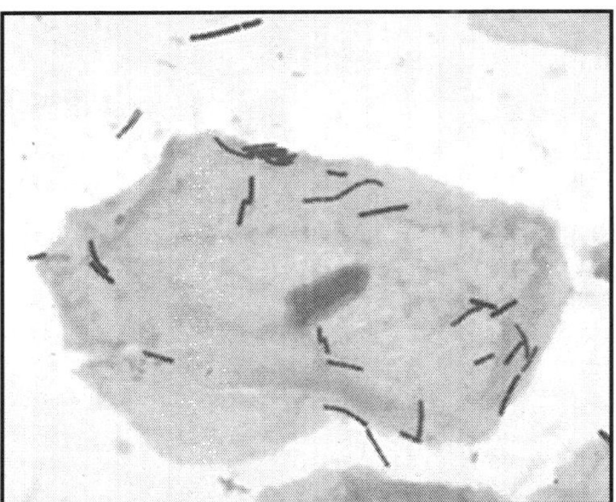

Fig. Lactobacilli in Association with a Vaginal Epithelial cell (CDC).

The bacteria are provided habitat with a constant temperature and supply of nutrients (glycogen) in exchange for the production of lactic acid, which protects the vagina from colonisation and disease caused by yeast and other potentially harmful microbes.

- *Commensalism:* There is no apparent benefit or harm to either member of the association. A problem with commensal relationships is that if you look at one long enough and hard enough, you often discover that at least one member is being helped or harmed during the association. Consider our relationship with *Staphylococcus epidermidis*, a consistent inhabitant of the skin of humans. Probably, the bacterium produces lactic acid that protects the skin from colonisation by harmful microbes that are less acid tolerant. But it has been suggested that other metabolites that are produced by the bacteria are an important cause of body odors (good or bad, depending on your personal point of view) and possibly associated with certain skin cancers. "Commensalism" best works when the relationship between two organisms is unknown and not obvious.
- *Parasitism*: In biology, the term parasite refers to an organism that grows, feeds and is sheltered on or in a different organism while contributing nothing to the survival of its host. In microbiology, the mode of existence of a parasite implies that the parasite is capable of causing damage to the host. This type of a symbiotic association draws our attention because a parasite may become pathogenic if the damage to the host results in disease. Some parasitic bacteria live as normal flora of humans while waiting for an opportunity to cause disease. Other non-indigenous parasites generally always cause disease if they associate with a non-immune host.
- Parasitology, actually a branch of microbiology, refers to the scientific study of parasitism but somehow it developed into a discipline that deals with eucaryotic parasites exclusively.

BACTERIAL STRUCTURE

The basic structural elements of a bacterial cell are represented schematically below.

Cell Wall

The cell wall consists of two or three layers of material, primarily peptide glycan. The cell wall mantains the shape of the bacterial cell and provides some protection from the external environment.

Cell Membrane

Closely applied to the inside of the cell wall is the cytoplasmic membrane, consisting of phosphatides, proteins and sugars.

Cytoplasma

The cytoplasm houses the metabolic apparatus of the cell and so contains enzymes, coenzymes, ribosomes and plasmids.

Nucleoid

The nucleoid contains the genetic material of the cell, in the form of nucleoproteins including DNA. The nucleoid does not have an enclosing nuclear membrane and this distinguishes bacteria from the eukaryotes.

Capsule

The capsule is an adhesive layer of material outside of the cell wall. The capsule is important in protecting bacteria against phagocytosis and lysis.

Flagellum

The flagellum provides the bacterial cell with motility.

Pilus

The pilus participates in the adhesion of bacteria to their growth substate, including other living cells.

- *Plasmid*: The plasmid is a circular DNA molecule, which can replicate independently in the cytoplasma.

GRAM-POSITIVE COCCI

Everyone has heard the trivial terms, "staph and strep germs." Use of the term, "germ," is a way of trivializing the idea of an infectious biological agent for children. We aren't quite sure why that sort of trivialization is continued into adulthood. Click for ideas on teaching bacteriology.

Staphylococcus Aureus

Staphylococcus is a very well known genus of bacteria. Colonies are "gold," or yellow on sheep blood agar solid media, hence the name. A common pathogen, boils, acne, wound infections, food poisoning are among a host of conditions caused by this organism. The organism is both pathogenic and invasive. It produces leukotoxin which can kill white blood cells and a wide variety of other toxins. *S. aureus* is quite pyogenic and in decades past was named Staphylococcus pyogenes, however that specific name is currently applied to one GPC,*Streptococcus pyogenes*. Increasingly, and especially in hospital, strains of both *S. aureus* and *S. epidermidis* have become resistant to the antibiotic DOC, methicillin. Such strains have been labelled Staph MR. Other clinically significant species include, *S. haemolyticus, S. hominis* and *S. saprophyticus*.

Staphylococcus sp. (Coagulase-negative) *Staphylococcus epidermidis*, appears as white colonies on sheep blood agar plates. While both *S aureus* and *S epidermidis* are normal inhabitants of the skin, epidermidis had been considered the lesser of the two in virulence. Most *S aureus* strains test positive for coagulase, clot formation in tube of citrated rabbit plasma. The other species test negative. "Infections by *S. epidermidis, S. haemolyticus*, and *S. hominis*, are associated with intravascular devices (prosthetic heart valves and intra-arterial or intravenous lines) and shunts. Also quite common are infections of prosthetic joints, wound infections, osteomyelitis associated with foreign bodies, and endocarditis." Dr GE Kaiser

Micrococcus sp. is a contaminant virtually every time. One definite exception is when it arrives in hospital accreditation test samples in which it is a "freebie." So it deserves mention. If you have ever looked at it under the microscope, you might wonder about the Micro- part of the name, since on gram stain, the cocci can be quite large. Pairs and clusters are seen. Catalase positive. Growth on mannitol agar. What else. "Three common species of *Micrococcus* are *M. luteus, M. roseus,* and *M. varians*."

Family *Streptococcaceae*

Streptococcus, plural streptococci, any member of a genus (*Streptococcus*) of spheroidal bacteria in the family *Streptococcaceae*. The term streptococcus ("twisted berry") refers to the bacteria's characteristic grouping in chains resembling a string of beads." "Streptococci can also be classified by the type of carbohydrate contained in the cell wall, a system called the Lancefield classification." Encyclopedia Britannica Agglutination and immunofluorescent antibody microscopic methods can quickly identify Lancefield groups.

Streptococcus Pyogenes

Streptococcus pyogenes electron micrographs, Rockefeller University, A downloadable set of eleven electron micrographs of*Streptococcus pyogenes* of exceptional quality.Click link or thumbnail: rockefeller.edu/vaf/ems.htm

Lancefield Group A, *S. pyogenes* species of streptococci cause rheumatic fever, scarlet fever, erysipelas, strep throat, tonsillitis, and other upper respiratory infections. In case you are wondering which streptococcal species is the worst, *Streptococcus pyogenes*, Lancefield group A is the bad boy in the Strep family. It is so important that physicians don't want any silly mistakes with names. "Pyogenes? What's that?" Physicians are programmed to react to the term "Strep group A."

A recently sensationalized strain of an especially invasive Lancefield group A *Streptococcus* carries the terms "pyo-," pus. "-gen-," forming. "necrotizing fascitis," flesh eating, to a very rare extreme. Once established under the skin, a focus of infection can spread liquifying flesh as it moves outwardly at a rate as rapid as one inch per hour. Sounds scary. You have five times more probability of getting hit by lightning. In common practice *Streptococcus* group A is usually found in samples from the throat, nasopharanyx, or in sputum in which a plethora of unimportant normal flora is also present. Its characteristic type of hemolysis is key for detection. If there is an area of clear or "beta," hemolysis, the possibility of the presence of group A organisms exists, although the same type of hemolysis pattern may also be exhibited by some other organisms such as some gram negative rods which may also be present. The bacterial colony is carefully scooped up, placed on a glass slide and mixed with a fluorescent antibody. If the organisms fluoresce when examined by fluorescent microscope, the presence of *Streptococcus* group A is confirmed. In laboratories which lack the relatively high tech fluorescent microscopes, the colonies are replated and streaked for isolation and identified with older methods. In these cases an extra incubation period, (day) is required.

Streptococcus Pneumoniae is a member of the Viridans group, so called because of green or, "alpha," hemolysis on sheep blood agar, causes lower respiratory infection pneumonia and upper respiratory infections bronchitis, laryngitis, otitis media (middle ear) and sinusitis.

The encapsulated, gram-positive coccoid bacteria have a distinctive morphology on gram stain, the so-called, "lancet shape," which actually looks more like a blunt arrow head.

"Before the 1990s, *Haemophilus influenzae* type b (Hib) was the leading cause of bacterial meningitis, but new vaccines being given to all children as part of their routine immunizations have reduced the occurrence of invasive disease due to H. influenzae. Today, *Streptococcus pneumoniae* and *Neisseria meningitidis* are the leading causes of bacterial meningitis."

"High fever, headache, and stiff neck are common symptoms of meningitis in anyone over the age of 2 years." "The diagnosis is usually made by growing bacteria from a sample of spinal fluid. The spinal fluid is obtained by performing a spinal tap, in which a needle is inserted into an area in the lower back where fluid in the spinal canal is readily accessible. Identification of the type of bacteria responsible is important for selection of correct antibiotics."

GBS is the most common cause of sepsis (blood infection) and meningitis (infection of the fluid and lining surrounding the brain) in newborns. GBS is a frequent cause of newborn pneumonia and is more common than other, better known, newborn problems such as rubella, congenital syphilis, and spina bifida. Before prevention methods were widely used, approximately 8,000 babies in the United States would get GBS disease each year. One of every 20 babies with GBS disease dies from infection. Babies that survive, particularly those who have meningitis, may have long-term problems, such as hearing or vision loss or learning disabilities.""Approximately one of every 100 to 200 babies whose mothers carry GBS develop signs and symptoms of GBS disease."" In pregnant women, GBS can cause bladder infections, womb infections (amnionitis, endometritis), and stillbirth. Among men and among women who are not pregnant, the most common diseases caused by GBS are blood infections, skin or soft tissue infections, and pneumonia. Approximately 20% of men and nonpregnant women with GBS disease die of the disease." "Many people carry GBS in their bodies but do not become ill. These people are considered to be "carriers." Adults can carry GBS in the bowel, vagina, bladder, or throat. People who carry GBS typically do so temporarily — that is, they do not become lifelong carriers of the bacteria." "GBS disease is diagnosed when the bacterium is grown from cultures of sterile body fluids, such as blood or spinal fluid. Cultures take a few days to complete. GBS infections in both newborns and adults are usually treated with antibiotics (e.g., penicillin or ampicillin) given through a vein."

"Group B streptococcus (GBS) is a type of bacterium that causes illness in newborn babies, pregnant women, the elderly, and adults with other illnesses such as diabetes or liver disease. GBS is the most common cause of life-threatening infections in newborns." *Streptococcus mutans,* Normal mouth flora. Is responsible for cavities. A vaccine is thought possible for *Streptococcus mutans* for the prevention of cavities. *Peptostreptococcus* sp. is anaerobic. It is sometimes opportunistically involved in the infection of wounds.

Gram-positive cocci (GPC) are seen on a Gram stain. If they occur primarily as mostly spherical cells arranged in grape-like clusters, then they can be reported as resembling staphylococci. On the other hand, if they occur primarily as somewhat elongated cells arranged in pairs or chains, then they can be reported as resembling streptococci/enterococci. Furthermore, the pneumococcus typically occurs as lancet-shaped pairs, often seen in the sputa of patients with bacterial pneumonia in the presence of high numbers of polymorphonuclear (PMNs, or "polys") white blood cells; such a direct Gram stain is considered diagnostic for the pneumococcus.

STAPHYLOCOCCI SPP

Staphylococci grow as large gray or white to yellow colonies on sheep blood agar. They show more robust growth under areobic than anaerobic conditions. They may be beta- or non-hemolytic. Unlike the streptococci and enterococci, staphylococci are catalase positive.

Colonies of *S. aureus* are frequently yellow (colour enhanced on chocolate agar) and beta-hemolytic. Coagulase-negative staphylococci (CNS) colonies are frequently gray to white and non-hemolytic. A trained microbiologist usually can discriminate *S. aureus* from CNS by colony morphology alone prior to performing any testing, but to prove the identification a coagulation test (slide coagulation and/or tube coagulation, depending upon the specimen source) is performed. *S. aureus* is coagulase-positive; CNS is coagulase-negative, within generally acceptable limits.

If CNS is recovered from the urine of a female, susceptibility testing against novobiocin may be performed. Novobiocin-sensitivity excludes the possbility of the isolate being *S. saprophyticus*, while resistance to novobiocin is consistent with an identification of *S. saprophyticus*. As with Group A *Streptococcus pyogenes*, *Staphylococcus aureus* is a very virulent organism (even though it can be found as normal skin flora), and in most cases when it is speciated it should be reported along with susceptibility test results.

STREPTOCOCCUS SPP

Streptococci grow as small gray colonies on sheep blood agar, where they may exhibit alpha- or beta-hemolysis, or they may be non-hemolytic. In contrast to the staphylococci, the streptococci show more robust growth under anaerobic than aerobic conditions and are catalase negative (the enterococci are catalase negative or weakly positive). The streptococci share many characteristics with the enterococci including the catalase reaction, but a trained microbiologist usually has little trouble distinguishing between the phenotypic characteristics of the two genera.

The first step in identifying streptococci is to make careful observation of growth characteristics on sheep blood agar:

- Small, strong alpha-hemolytic colony, especially from a pulmonary or oropharyngeal specimen, or from blood or cerebrospinal fluid, otherwise found as oral contamination: consider *S. pneumoniae* or viridans streptococci (*S. pneumoniae* frequently will have undergone some degree of autolysis, producing a typical "donut colony").
- Medium gray colony with strong or weak alpha-hemolysis: consider especially *Enterococcus*.
- Medium gray non-hemolytic colony, most commonly seen in mixed respiratory cultures or in low numbers in a urine culture: consider especially non-hemolytic streptococci.
- Small gray beta-hemolytic colony with small to medium zone of beta-hemolysis: consider especially Group A *S. pyogenes*.
- Medium gray colony with very small zone of weak beta-hemolysis, perhaps limited to hemolysis occurring only directly under the colony; colony thins gradually at the periphery producing a "hazy beta" colony, especially from a female urogenital site, a newborn blood culture, and occasionally from wound cultures: consider especially Group B *S. agalactiae*or *L. monocytogenes*.
- Small to medium gray colony with large zone of strong beta-hemolysis, especially from a throat culture: consider especially beta-hemolytic streptococci not Group A and not Group B.

Gray alpha-hemolytic colonies may be suspected of being *S. pneumoniae*, viridans streptococci, or *Enterococcus* sp. Of these three taxa, only the enterococci will produce a positive PYR test (red colour produced after addition on N,N methyl aminocynnamaldehyde reagent after exposure to L-pyrrolidonyl-beta-naphthylamide (PYR) substrate); both *S. pneumoniae* and the viridans streptococci are PYR-negative. The most rapid test to discriminate between *S. pneumoniae* and viridans streptococci is bile solubility, in which a single drop of 40% sodium desoxycholate (a bile salt) is dropped onto a single, well-isolated colony of the alpha-hemolytic organism. After about 15 minutes the plate is re-examined. The autolytic system of *S. pneumoniae* is activated by the bile salt and the colony will have dissolved, whereas the viridans streptococci are bile-insoluble, and the colony will remain intact on the plate. An alpha-hemolytic colony whose morphology resembles that of *Enterococcus* more than it resembles the viridans streptococci or the pneumococcus but which is PYR negative may be a Group D *Streptococcus*. Alpha-hemolytic Group D streptococci are sometimes encountered as significant pathogens urine cultures, although they may also (rarely) cause endocarditis and/or septicemia. Like the enterococci, Group D streptococci are able to grow in the presence of 40% bile esculin and to hydrolyze esculin, while other streptococci are not, and can be identified by these tests.

The viridans streptococci are generally of low virulence, although they are found in about 50% of cases of subacute bacterial endocarditis, particularly in immunocompromised patients after a transient bacteremic event such as orthodontal manipulation.

Some microbiologists consider the non-hemolytic streptococci to be part of the viridans group, although "viridans" refers to the greening of blood agar. Non-hemolytic streptococci are usually found as non-pathogenic contaminants or as part of a mixed culture, including after a transient bacteremic event such as orthodontal manipulation. The finding of an apparent non-hemolytic streptococcus should prompt the microbiologist to exclude enterococci (which are PYR positive), Group D streptococci (which grow in the presence of 40% bile and which hydrolyze esculin), weakly-hemolytic group B *S. agalactiae* (consider especially colony morphology, specimen type, numbers of organism present, and perhaps a specific test such as latex agglutination for Group B *S. agalactiae*), and *L. monocytogenes* (colony morphology nearly identical to that of Group B *S. agalactiae*, but a catalase-positive Gram-positive rod). To some extent the approach to identifying beta-hemolytic streptococci depends upon the specimen type.

Group A *S. pyogenes* is the most important streptococcal human pathogen. It produces a small to medium gray colony with a small zone of beta-hemolysis and is most commonly recovered from a throat culture, where it produces pharyngitis and tonsillitis most frequently in children aged 5-15 years. Asymptomatic carriers in the upper respiratory tract and on the skin are often responsible for the spread of infection. It is important to diagnose and treat "strep throat" not so much because of the pharyngitis per se but because of the possible sequelae including rheumatic fever and acute glomerulonephritis. Although other streptococci may be associated with pharyngitis they are not associated with these sequelae, and antibiotic treatment is therefore not indicated in the absence of a positive finding of Group A *S. pyogenes*, since the possibility of the emergence of antimicrobial resistance outweighs any therapeutic benefit.

Group A *S. pyogenes* recovered from any other site such as wound, fluid, or blood cultures should be considered to be a serious pathogenic threat since the organism can be highly virulent and rapidly lethal. (Occasionally Group A *S. pyogenes* is found in low numbers in urine cultures, especially in the presence of mixed flora, when it is probably present as normal flora, although mention of its presence even in low numbers may be indicated.)

Group B *S. agalactiae* produces a medium gray colony with a small to very small zone of weak beta-hemolysis which sometimes may be observed to occur only directly underneath the colony after the colony has been physically removed from the blood agar plate. The colony thins gradually at the periphery producing a typical "hazy beta" colonial morphology. This organism is recovered especially from female urogenital sites (including, not uncommonly, urine cultures), from newborn blood cultures, and occasionally from wound cultures. About a third of women are asymptomatic vaginal carriers of Group B *S. agalactiae*, but the organism can cause severe neonatal infection including pneumonia, meningitis, and septicemia of the newborn. Wound infections with Group B*S. agalactiae* are sometimes encountered, sometimes in diabetic patients. On the basis of colony morphology alone, Group B *S. agalactiae* can be confused with the catalase-positive, Gram positive bacillus *Listeria monocytogenes*. Non-Group A, non-Group B beta-hemolytic steptococci produce small to medium gray colonies with large zones of strong beta-hemolysis and are recovered especially from oropharyngeal cultures, where they are rarely confused with *S. pyogenes*. They are of minimal virulence.

ENTEROCOCCI SPP

Enterococci grow as small to medium gray colonies on sheep blood agar. Like the streptococci, they show more robust growth under anaerobic than aerobic conditions. They may be alpha-, beta-, or non-hemolytic. They are catalase negative or weakly positive, unlike the staphylococci, which are vigourously catalase positive. A weakly positive catalase reaction may be as little as a tiny stream of bubbles after a short delay of time.

The enterococci have always been dogged by controversy in medical microbiology, and they have become even more controversial recently since:

- They were taxonomically separated from the streptococci when the genus *Enterococcus* was created in 1984, and

- They first demonstrated resistance to vancomycin in 1986.

Before 1984 the enterococci still were in the genus *Streptococcus*, although it had long been appreciated that their ecology differed significantly from that of other streptococci; specifically, enterococci are found as normal flora in the large bowel, while the streptococci are found more commonly as oropharyngeal organisms. Indeed, in many respects the enterococci are best thought of as being similar in ecological aspects to the enteric Gram-negative rods, although the enterococci just happen to be Gram-positive cocci. Still, from a clinical microbiologist's perspective, they appear similar to streptococci in both Gram stain and colonial morphology. They are catalase-negative or (more commonly) weakly positive, and they can be alpha-, beta- or non-hemolytic on sheep blood agar (by far most human isolates are weakly alpha-hemolytic or non-hemolytic, and beta-hemolytic enterococci are very rarely encountered). Also, like Group D *Streptococcus* spp., the enterococci can grow in the presence of 40% bile salts and hydrolyze esculin, and they express the Lancefield Group D antigen; unlike Group D *Streptococcus* spp. the enterococci are PYR-positive. For these reasons the enterococci are identified using the same protocol as the streptococci in the clinical microbiology laboratory, but it should never be forgotten that ecologically their behaviour is much closer to enteric Gram-negative rods than it is to the streptococci.

GRAM POSITIVE BACILLI

Anaerobic gram-positive bacilli are more well known than some of their aerobic counterparts; everybody in the known world has heard of the Clostridia. There were only a few genera of them so we decided to begin with them. *Actinomyces* sp. Filamentous gram positive obligate anaerobic rods cause actinomycosis. The most common species is *A. israeli.* Infection is usually the result of trauma. Pus and fluid at the site of infection have characteristic sulfur granules. DOC: penicillin or tetracycline.

Bifidobacterium are bone shaped gram positive obligate anaerobic opportunistic pathogens which are normal inhabitants of the gut. The ends of the rods show characteristic splitting on gram stain and the organism shows branching filaments as well. *B. dentium* is the most common species.

Clostridium sp

The University of Wisconsin online Textbook of Microbiology has an excellent section on the clostridia. In fact, there is an excellent gram stain image of pus from a mixed anaerobic infection. Gram stains are the, "real world of clinical microbiology," and represent the routine view of what clinicians actually see of bacterial organisms. In contrast EM images are more popularly viewed by laypersons. The following quote lists the general characteistics of clostridia.

The clostridia are relatively large, Gram-positive, rod-shaped bacteria. All species form endospores and have a strictly fermentative mode of metabolism. Most clostridia will not grow under aerobic conditions and vegetative cells are killed by exposure to O2, but their spores are able to survive long periods of exposure to air. The clostridia are ancient organisms that live in virtually all of the anaerobic habitats of nature where organic compounds are present, including soils, aquatic sediments and the intestinal tracts of animals."

Clostridium Botulinum

Everybody knows that you have to be careful when canning meat, fish, fruits and veggies so that you don't get botulism. Didn't your grandmother ever tell you that botulism is the world's worst poison, so you have to be very careful? Well, guess what? It is. Don't even taste foods suspected to be spoiled by clostridium botulinum. But just to get your ducks in a row; *botulinum* is the name. Botulism is the condition. Botulin is the poison itself.

People who work in clinical bacteriology are very unlikely to see this bacterium unless:

- You work in a bacteria bank where they keep lyophilized specimens of every strain of bacteria.
- You are doing epidemiology work for the CDC and are trying to locate the source of an outbreak of botulism poisoning.
- You are doing research work involving gene sequencing of *C botulinum* strains.

According to the CDC, "In the United States an average of 110 cases of botulism are reported each year. Of these, approximately 25% are foodborne, 72% are infant botulism, and the rest are wound botulism. Outbreaks of foodborne botulism involving two or more persons occur most years and usually caused by eating contaminated home-canned foods.

Clostridium difficile They probably named them, "difficile," because they are difficult. Or possibly because of the difficulty in remebering that it causes, "pseudomembranous colitis." Probably some people have known of someone who had strong antimicrobial therapy to clean out their intestines for one reason or another who ended up with a nasty form of enteritis caused by *C. difficile* which wasn't killed off by the antibiotics on the first go around. The doctor didn't even have to do a culture, he knew what had happened, didn't he? When he performed an endoscopy even the patient could see all those big white spots. He zapped those bugs on the second go around though. What a relief. It's like hitting yourself on the head with a hammer. It feels so good when it stops. This little difficulty with *Clostridium difficile* occurs in Americans, Europeans and Japanese. *C.*

Clostridium perfringens is an invasive pathogen which posesses a, "huge array of invasins and toxins, causes wound and surgical infections and also severe uterine infections, especially when coat hangers are used for abortions. Old timers remember someone who got gas gangrene while living in the trenches in WWl. Some called it foot rot, or boot rot. The foot and leg swell up due to the gas produced and if you press down on it, the gas bubbles out. It smells really fetid like the putrid smell of rotted meat. Guess what? It *is* rottet meat. *Your* rotted meat.

The Clostridia which includes *C. perfringens* are at the bottom of the food chain. They live in soil and they breakdown not only the proteins but, the amino acids. When they break down lysine and arginine, the straight chain amines putrecine and cadaverine are produced. These small volatile molecules float through the air and when they reach the receptors in the nostrils, they elicit an electrical impulse which travels along a special pathway to a particular part of the brain. Over millions of years humans who have reacted to the resulting sensation of a, "stinky," smell by assiduously avoiding its source have survived by not being breakfast for these organisms.

Clostridium tetani Everybody knows that if you cut yourself with a metal object or get dirt in a cut, you have to get a nasty tetanus shot. A lot of people know that Clostridia form spores but, they can't remember which Clostridia form which kind of spores. Here's a hint, "T" is for terminal and also tetani, get it? It's a little demonic mnemonic to help you remember on your medical micro proficiency exam.

PROTEOBACTERIA IN PHYLUM

This is the most diverse and enormous of all bacterial groups. All are Gram Negative and their 16s rRNA nucleotide sequences have a high percentage in common. There are 5 Proteobacteria classes, Alpha, Beta, Gamma, Delta and Epsilon.

Alphaproteobactria

These are aerobes. They grow at low nutrient levels. Some are pleomorphic while others are rods to curved rods as well as cocci and spirals. Most species have prosthecae, a cytoplasmic extension which is surrounded by a cytoplasmic membrane and a cell wall. The prosthecae is for attachment and better nutrient absorption.The pathogenic Alphaproteobacteria are the Rickettsia, Ehrlichia and the Brucella genus. The Rickettsia causes Rocky Mt. Spotted Fever. The Brucella causes abortions and sterility in animals The Ehrlichia causes Ehrlichiosis, which is a tick borne disease, where the organism lives within the white blood cells.

Betaproteobacteria

Also grows well in low nutrient level areas. The main difference in the Alpha vs the Beta is their rRNA sequences.

- *Example*: Nitrosomonas – a soil nitrifying microbe, by oxidizing Nitrite to Nitrate. The pathogenic Betaproteobacteria are the genus Neisseria. These are Gram Negative Diplococci bacterium. They cause PID, Gonorrhea, meningitis and cervix inflammation. Another genus is the genus Bordetella. This organism causes Pertussis (Whooping cough).

Gammaproteobacteria

Of all the Proteobacteria, this is the largest group.
Intracellular Pathogens of the Gammaproteobacteria:

- *Legionella:* A human pathogen.
- Can live inside immune cells.
- Get their energy from the metabolism of amino acids inside the WBCs.
- Legionella causes Legionnaires Disease, a type of Pneumonia.
- *Coxiella*: A human pathogen.
- Can live inside the phagolysomes of WBCs.
- Grows well within an acid pH environment.

Deltaproteobacteria

This is a smaller group of gram negative bacteria. One main example is the genus Desulfovibrio. These organisms can be found in the sediments of a polluted stream and in sewage treatment lagoons. In these type areas, this bacterium will release H_2S as an end product of its anaerobic respiration.

This H_2S by-product will produce FeS (iron-sulfide) when it reacts with Iron. Since these are sulfate reducing bacteria, they are the ones that will corrode the pipes in sewer lines and heating systems. Another example is the Bdellovibrio. This is not a human pathogen, but it will attack G. Neg bacteria. This organism will collide with a GN bacterium and bore through its cell wall and reside within the periplasmic space. It will kill off the GN bacterium by impeding protein synthesis, DNA and RNA synthesis as well as disrupting the cytoplasmic membrane structure. As the GN organism begins dying, the Bdellovibrio feeds on its nutrients and grows into a long filament. The filament breaks into smaller cells when it is released from the dead GN bacterium. The new cells produce a flagellum and swim off to repeat the process within another GN organism.

Epsilonproteobacteria

Morphological structure can be spiral, GN rod or vibrio shape. The two genus of this group are Campylobacter and Helicobacter. Campylobacter may inflame the intestinal tract as well as cause blood poisoning. Helicobacter is the causing agent of stomach ulcers.

Other Gram Negative Bacteria

Chlamydia

Gram negative cocci that are small in size. Only within the cells of birds, mammals and some invertebrates will this organism replicate. The smallest Chlamydias is smaller than the largest viruses, which is the Poxvirus. They contain both DNA and RNA, but lack peptidoglycan in the cell wall. This bacterium is an Obligate Intracellular Parasite. This is the most sexually transmitted bacteria in America today.

Once the Chlamydia parasitizes a host cell, the initial body it forms is called a "reticulate body". The initial body has numerous binary fission divisions. This fills the cell with reticulate bodies. These reticulate bodies now turn into "elementary bodies". This body is the infective stage and it is very resistant to drying out. The Elementary Body is released when the host cell dies. The cell now drifts until it encounters another host cell. After entering the host cell, the cycle repeats its self.

The Spirochetes

These are helical bacteria. They move by axial filaments. These filaments are located within the periplasmic space. When the filaments rotate, the organism moves through its environment with a corkscrew motion. Treponema pallidum is one spirochete that causes Syphilis. Borrelia burgdorferi will cause Lymes Disease.

The Bacteriods

These are obligate anaerobes. Normal flora of animals and human digestive tracts. This bacterium helps to digest substances that are hard for these two organisms to digest, like cellulose. 30% of the flora in human fecal matter is this genus. This is the most common anaerobic human pathogen. The genus Cytophaga is an aquatic aerobe. They can degrade complex carbohydrates like cellulose, chitin, agar and pectin. Their action is detrimental to wooden boats and dock piers. They are important is raw sewage degradation.

The Protozoa

Protozoans have three characteristics:

1. Cell wall is absent
2. Unicellular
3. Eukaryotic.

They move by either Pseudopodia (false feet), Flagella or Cilia.

The Amoebae

- Found in Phylum Rhizopoda

- These organisms lack mitochondria.
- All reproduce by binary fission.
- Found and live in the soil, freshwater and seawater.
- Primary food is Red Blood Cells.
- Transmission between humans is the fecal-oral route.

Entamoeba histolytica causes dysentery and/or intestinal ulceration. This is the only pathogenic amoeba found in human intestines. Nagleria fowleri can cause brain abscesses as well as the disease Primary Amoebic Meningioencephalitis. Transmission is by contaminated lake water entering the nasal cavities and migrating to the brain. 99+% fatal and this disease is diagnosed by autopsy.

The Ciliates

- Found in Phylum Ciliophora.
- Has two nucleus.
- The Macronucleus contains the genetic material.
- It also controls growth and metabolism.
- The Micronucleus is used for reproduction by conjugation.
- The Paramecium is the most well known example.
- Balantidium coli is the only pathogenic ciliate in humans.
- Didinium, the ciliate that will phagocytize other protozoans.
- Vorticella creates a whirlpool motion to direct food into its mouth.

The Flagellates

- Found in Phylum Zoomastigophora.
- These are eukaryotes that lack mitochondria.
- Leishmania, causes Leishmaniasis, which are disfiguring ulcers that distort the human body.
- Trichomonas is an STD.
- Trypanosoma causes African Sleeping Sickness.
- Giardia causes diarrhea from ingesting contaminated water.

The Dinoflagellates

- Photosynthetic, unicellular microbes.
- Have cell walls composed of cellulose.
- Their food reserves are oil and starches.
- Many have Silica protective armor shields.
- These organisms are historically classified as Algae.
- Their chromosomes lack histone proteins.
- The motile organisms have two flagella.
- The transverse flagella wraps around inside the middle groove of the organism and its beating motion causes the cell to spin.

- The posterior flagella extends away from the cell and helps it to move forward.
- Many of these are bioluminescent.
- The toxin is called Saxitoxin.
- The Dinoflagellates that produce a red pigment will come together in a body of water and create a "Red Tide".
- The genus Gymnodinium and Gonyaulax produce neurotoxins.
- These toxins act on the nervous system.
- If shellfish eat the Dinoflagellate and the shellfish is then consumed by humans, causing Paralytic Shellfish Poisoning.
- Pfiesteria, another Dinoflagellate that produces a toxin.
- A person can contact this toxin by handling fish that are infected with the toxin or you only have to breathe air that is laden with the microbes.
- This Pfiesteria toxin can cause memory loss, along with muscle cramps, diarrhea, vomiting and nausea.

If the Dinoflagellate Gambierdiscus is consumed by large fish, it will produce the toxin called Ciguatera. Humans get the toxin directly from eating the contaminated fish. If Mussels eat certain Diatoms, the diatom will produce a toxin called Domoic Acid. The biggest side effect is Memory Loss.

FUNGI

- Kingdom Fungi includes the molds, yeasts and mushrooms.
- These organisms are chemoheterotrophic.
- They have cell walls that are composed of chitin.
- They do not exhibit photosynthesis because they have no chlorophyll.

FUNGI SIGNIFICANCE

- Fungi recycle the nutrients from decomposing dead plant and animals.
- They also form mycorrhizae, which is an association with plants and helps them to absorb dissolved minerals and water.
- Fungi are used by humans for food and beverages.
- The mold genus Penicillium synthesizes Penicillin and Cephalosporium synthesizes the antibiotic called Cephalosporin.
- Molds also synthesize Cyclosporin, the drug used in helping a person not to reject an organ transplant.
- The chemical called Mevinic acid, which is used to reduce a persons total cholesterol level is synthesized by molds.
- The yeast Saccharomyces cerevisiae, which is Bakers Yeast, was the first eukaryotic cells to have its total genome sequenced.
- This organism is used widely in Biotechnology.

- Approximately 30% of all fungi can cause disease in humans, plants and animals called Mycoses.
- Fungi are very resilient, by tolerating varying conditions of sugar, salt and acid.
- Foods exposed to the environment, like jams, jellies, fruit and pickles can be contaminated by Fungi.

FUNGAL MORPHOLOGY

- A Thallus is the term given to a fungal body.
- The Thalli of a mold consist of long, tubular branches filaments called Hyphae.
- Hyphae can be either Septate or Aseptate.
- The Septate hyphae have septum or cross walls.
- The Aseptate hyphae have no cross walls and can be called Coenocytic.
- Temperature also has effects on Fungi.
- If a fungus grows at 25°C or ambient temperature, it will be in mold form and produce moldlike thalli.
- If the fungus grows at body temperature of 37°C, it will be in yeast form and will produce yeastlike thalli.
- A fungus that will take on two different form because of the influence of temperature, are called Thermal Dimorphic.
- When a mold grows on a medium and produces its tangled mass of hyphae, this collective mass is termed a Mycelium.

FUNGI NUTRITION

- Absorption is the mechanism by which a Fungus acquires its nutrients.
- Fungi (like Flies) secrete enzymes from inside their bodies on to the nutrient source.
- The enzymes catabolized the organic nutrient and degrade it into its smaller component parts.
- Then the smaller nutrient is transported into the thalli.
- This means Fungi are saprobes or saprophytic.
- Dead plants or animals are their nutrient source.
- Some Fungi will take nutrients from animals or plants that are alive.
- These Fungi have hyphae that are modified and are called haustoria.
- The Fungi will inject its haustoria into the nutrient host cell and pull the nutrients out.

FUNGI CLASSIFICATION

- Division Zygomycota
- These are the coenocytic molds.

- Some are saprobes, while the others are other fungi and insect obligate parasites.
- Example is the Rhizopus nigricans, the Black Bread Mold.
- Division Ascomycota
- These are the Fungi that spoil food.
- They form haploid ascospores inside sacs called asci.
- Claviceps purpurea will grow on grain products and synthesize a chemical called lysergic acid diethylamide (LSD), which causes humans to hallucinate and causes cattle to have abortions.
- The Ascomycete Penicillium is our main source for the antibiotic called Penicillin.
- Another Ascomycete called Saccharomyces cerevisiae, called Bakers Yeast is the mainstay for the brewing and baking industry.

Division Basidiomycota

This type fungi is world wide and can be found in the forms of jelly fungi, mushrooms, bracket fungi, puffballs, birds nest fungi and stinkhorns. Toadstools is the name sometimes given to poisonous mushrooms. The Basidiomycetes produce fruiting bodies or mushrooms called Basidiocarps. Fungal meningitis is caused by the basidiomycete yeast Cryptococcus neoformans.

Division Deutreromycota

- These are terrestrial saprobes.
- They are also plant pathogens and well as fungi pathogens.
- No sexual stage has ever been observed with this Division.
- Trichophyton tonsurans is the deuteromycete that can cause "Ring Worm".
- Histoplasmosis is another fungal disease caused by Histoplasma capsulatum.
- This organism is also thermal dimorphic.

ALGAE

- Phycology is the study of Algae.
- They live in extreme environments from soil to ice.
- In water, they live in the photic zone.
- This is the zone of water that is penetrated by sunlight.
- Blue wavelengths penetrate only about 1 metre into the surface of water.
- Algaes main photosynthetic pigment is chlorophyll a.
- This pigment captures red light.
- If algae is going to survive in deeper waters, it must have accessory pigments for photosynthesis.

- The Red Algae contains the red pigment phycoerythrin, which absorbs
- Blue light and this allows the red algae to grow in deeper parts of the photic zone.
- Algae cell walls are composed of cellulose.
- They produce 50% of all the oxygen and carbohydrates that are synthesized.
- Free-floating microscopic algae are called Phytoplankton.
- Microscopic animals that feed on the phytoplankton are Zooplankton.

ARCHAE BACTERIA

Modern taxonomists have placed all organisms into 3 domains.

- Archaea
- Bacteria
- Eukarya - the largest of the three groups.

These three domains are assigned because of their similarities in DNA, RNA or protein sequences.

The Archaea

This is a distinct type of prokaryotes. Their cells walls lack peptidoglycan. The lipids in their plasma membranes have highly branched hydrocarbon chains. There are two phyla of the Archaea: Crenarchaeota and Euryarchaeota. These are named so because of their rRNA sequences. Their cells walls are composed of either proteins, polysaccharides, glycoproteins or lipoproteins. None of this group have the ability to cause disease in animals or humans.

TYPES OF ARCHAEA ORGANISMS

The Extremeophiles require extreme conditions of either salt, pH or heat.

- *The Thermophiles*: Organisms that have to live about 45°C. are in the phylum Crenarchaeota.
- *Examples are*:
  - *Sulfolobus*: Live in Yellowstone National Park in the acid and sulfur-rich hot springs. Optimum pH is 2.Can live in terrestrial volcanic habitats.
  - *Pyrodictium*: Live in deep sea hydrothermal vents.

Their shape is irregular-disks with elongated proteins that help them attach to sulfur grains. The sulfur acts as their final electron acceptor in their respiration. A strict anaerobe. An optimum pH of 6.0.

The Acidophiles

- *Acidanus*: Live in environments with a lower than normal pH value. A facultative anaerobe with an optimum pH of 2.

- *The Halophiles*: Optimal salt concentration range is 17% - 23%. Phylum Euryarchaeota contains the halophiles. Live in the Dead Sea, the Great Salt Lake and solar evaporation ponds (salt for seasoning and fertilizer production).

They spoil pork, sausage and salted fish.They have an optimum pH, but still can live and grow from 9% to 35% salt. They produce red-pigments, which protects them against UV light.

The Halophile called Halobacterium salinarium is:

- The most studied halophile.
- Uses light energy to drive synthesis of atp.
- Lacks photosynthetic pigments.
- It synthesizes purple proteins called bacteriohodopsins.

They absorb light energy and use it to pump protons across the plasma membrane and set up a proton gradient for atp production. Able to rotate its flagella with energy supplied from the proton gradient and position its self at the optimum depth for light absorption. The Methanogens – are obligate anaerobes.

Found in the phylum Euryarchaeota. Convert CO_2, organic acids and H_2 into methane gas. This is the Archaeas largest group.Most species in this group are Mesophiles.

Halophilic Methanogens have been discovered. The genus Methanobacterium, lives in the large intestine of animals. This is the basic source of methane in the environment.

TYPES OF BACTERIA ORGANISMS

Phototrophic bacteria – produce organic compounds from CO_2.

There are five groups of these type bacteria:

1. Blue-green algae – cyanobacteria
2. Green sulfur bacteria
3. Green non-sulfur bacteria
4. Purple sulfur bacteria
5. Purple non-sulfur bacteria

Low G + C Gram Positive Bacteria

Taxonomists are now classifying microbes according to their Guanine to Cytosine ratios (G + C).If this ratio is below 50%, they are considered "low". Another characteristic is the similar sequences within their 16s rRNA. The Low G + C Gram Positive bacteria have similar sequences in their 16s r RNA, while the High G + C Gram Positive bacteria have similar sequences in their 16s rRNA. These Low G + C bacteria are in the Phylum Firmicutes. The Firmicutes includes Clostridia, Mycoplasmas and the Low G + C Gram Positive Bacteria.

The Clostridia

The genus Clostridium has a bacillus (rod) morphology and are obligate anaerobes. Most form endospores. This genus can produce toxins that cause disease in humans. The genus Epulopiscium can be seen without a microscope. Appears ciliated with its short flagella. Does not reproduce by binary fission, but by forming new cells within and releasing them through a slit in the parents cell wall.

Veillonella is a sulfate reducing microbe. Produces H_2S from elemental sulfur thru anaerobic respiration. Part of the biofilm which is attached to the teeth of warm-blooded animals (humans). It has a gram positive wall structure, but stains negatively!

The Mycoplasmas

These can be either obligate or facultative anaerobes. They have no cell wall and will stain pink, negatively. Their cytoplasmic membranes contain lipids called sterols.

This makes their membranes rigid and strong. No cell wall also means they are pleomorphic, which means they have no definite shape. These are the smallest metabolizing organisms known. Form a "fried egg" appearance when growing on solid media. This organism causes Mycoplasma pneumonia in humans.

The Genus Bacillus

A large Gram positive rod. Found commonly in the soil. Form endospores. Move by Peritrichous flagella. Are either aerobes or facultative anaerobes. Many species cause disease. Many species give us antibiotics. Most are laboratory contaminants.

The Listeria

A small Gram Positive rod. Main specie is (Listeria) monocytogenes. Will contaminate meat products and milk. Does not produce endospores. Grows well under refrigeration. Will survive within phagocytic WBCs. Is virulent when it crosses the placenta and infects the fetus. Can cause meningitis. Causes a bacteremia in those that are aged, with cancer, have diabetes, have AIDS or are immunosuppressed.

The Lactobacillus

Non-spore forming rods. Normal flora of the mouth, vagina, stomach and intestinal tract.Protect the body by microbial antagonism, which is inhibiting the growth of pathogens. Used commercially to make yogurt, buttermilk, sauerkraut and pickles.

The Enterococcus and Streptococcus

These are both gram positive cocci either in pairs or chains. Cause a lot of human illnesses. Some have become resistant to antimicrobial drugs.

The Staphylococcus

Staphylococcus aureus is found one human skin or nasal cavities. The strains that produce toxins can cause disease in humans. MRSA is a medically important species of Staphylococcus.

THE HIGH G + C GRAM POSITIVE BACTERIA

These High G + C Gram Positive bacteria are found in the Phylum called Actinobacteria. This phylum includes those organisms that are filamentous or rod shaped. The Bacteria of the High G + C Gram Positive Group.

The Corynebacterium

These organisms are pleomorphic. Rod shaped, aerobes and facultative anaerobes.Reproduce by snapping division, forming palisades and V-shapes. They are characterized by storing phosphate within their inclusions called metachromatic granules.

The Mycobacterium

These organisms replicate slowly and are aerobes. Slightly curved shaped rods. They secrete Mycolic acids to the outside of the cell and this slows their growth pattern..These acids make the cells resistant to staining and drying out.

- *Examples are*: Mycobacterium tuberculosis–causes T. B.Mycobacterium leprae – causes leprosy.
- The Actinomycetes type organisms:
- This organisms resemble fungi with branching filaments.
- The filaments are made up of prokaryotic cells.
- These filamentous bacteria usually only cause disease in those that are immunosuppressed.
- Three Actinomycetes genus are:
- Actinomyces – are a facultative Capneic filaments.(Capneic – grows well in high concentration of CO_2).
- Normal flora of human mouth and throat mucous membranes.
- *Example*: Actinomyces israelii–will infect and destroy human tissue by abscessing them and then migrate to the abdomen, where it causes organ dysfunction.
- *Nocardia*: Have subterranean and aerial filaments that resemble fungi.

Live in the soil and water niches. This is a pollutant degrading microbe of the environment. Will break down waxes, rubber, detergents, pesticides, petroleum hydrocarbons, benzene and PCBs (polychlorinated biphenyls). A few species causes lesions in humans.

- *Streptomyces*: Can degrade carbohydrates like lignin, chitin, cellulose, latex and keratin. When the soil smells musky, its by this organism breaking down material in the soil. Tetracycline is one antibiotic from this genus.

BACTERIAL PATHOGENESIS

A pathogen is a microorganism (or virus) that is able to produce disease. Pathogenicity is the ability of a microorganism to cause disease in another organism, namely the host for the pathogen. As implied above, pathogenicity may be a manifestation of a host-parasite interaction.

In humans, some of the normal bacterial flora (*e.g. Staphylococcus aureus, Streptococcus pneumoniae, Haemophilus influenzae)* are potential pathogens that live in a commensal or parasitic relationship without producing disease. They do not cause disease in their host unless they have an opportunity brought on by some compromise or weakness in the host's anatomical barriers, tissue resistance or immunity. Furthermore, the bacteria are in a position to be transmitted from one host to another, giving them additional opportunities to colonise or infect. There are some pathogens that do not associate with their host except in the case of disease. These bacteria may be thought of as obligate pathogens, even though some may rarely occur as normal flora, in asymptomatic or recovered carriers, or in some form where they cannot be eliminated by the host.

OPPORTUNISTIC PATHOGENS

Bacteria which cause a disease in a compromised host which typically would not occur in a healthy (non-compromised) host are acting as opportunistic pathogens. A member of the normal flora can such as *Staphylococcus aureus* or *E. coli* can cause an opportunistic infection, but so can an environmental organism such as *Pseudomonas aeruginosa.* When a member of the normal flora causes an infectious disease, it sometimes referred to as an endogenous bacterial disease, referring to a disease brought on by bacteria 'from within'. Classic opportunistic infections in humans are dental caries and periodontal disease caused by normal flora of the oral cavity.

A photomicrograph of *Pseudomonas aeruginosa*, one of the most common opportunistic pathogens of humans. The bacterium causes urinary tract infections, respiratory system infections, dermatitis, soft tissue infections, bacteremia and a variety of systemic infections, particularly in cancer and AIDS patients who are immunosuppressed. CDC.

INFECTION

The normal flora, as well as any "contaminating" bacteria from the environment, are all found on the body surfaces of the animal; the blood and internal tissues are sterile. If a bacterium, whether or not a component of the normal flora, breaches one of these surfaces, an infection is said to have occurred. Infection does not necessarily lead to infectious disease. In fact, infection probably rarely leads to infectious disease. Some bacteria rarely cause disease if they do infect; some bacteria will usually cause disease if they infect. But other factors, such as the route of entry, the number of infectious bacteria, and (most importantly) the status of the host defences, play a role in determining the outcome of infection.

DETERMINANTS OF VIRULENCE

Pathogenic bacteria are able to produce disease because they possess certain structural or biochemical or genetic traits that render them pathogenic or virulent. (The term virulence is best interpreted as referring to the degree of pathogenicity.) The sum of the characteristics that allow a given bacterium to produce disease are the pathogen's determinants of virulence.

Some pathogens may rely on a single determinant of virulence, such as toxin production, to cause damage to their host. Thus, bacteria such as *Clostridium tetani* and *Corynebacterium diphtheriae*, which have hardly any invasive characteristics, are able to produce disease, the symptoms of which depend on a single genetic trait in the bacteria: the ability to produce a toxin. Other pathogens, such as *Staphylococcus aureus*, *Streptococcus pyogenes* and *Pseudomonas aeruginosa*, maintain a large repertoire of virulence determinants and consequently are able to produce a more complete range of diseases that affect different tissues in their host.

A photomicrograph of *Corynebacterium diphtheriae* bacteria using a Gram stain technique. *Corynebacterium diphtheriae* causes diphtheria that affects the upper respiratory tract, where an inflammatory exudate causes severe obstruction to the breathing airways, and sometimes suffocation. CDC.

PROPERTIES OF THE HOST

The host in a host-parasite interaction is the animal that maintains the parasite. The host and parasite are in a dynamic interaction, the outcome of which depends upon the properties of the parasite and of the host. The bacterial parasite has its determinants of virulence that allow it to invade and damage the host and to resist the defences of the host. The host has various degrees of resistance to the parasite in the form of the host defences.

HOST DEFENCES

A healthy animal can defend itself against pathogens at different stages in the infectious disease process. The host defences may be of such a degree that infection can be prevented entirely. Or, if infection does occur, the defences may stop the process before disease is apparent. At other times, the defences that are necessary to defeat a pathogen may not be effective until infectious disease is well into progress.

Typically the host defence mechanisms are divided into two groups:

1. *Innate Defences:* Defences common to all healthy animals. These defences provide general protection against invasion by normal flora, or colonisation, infection, and infectious disease caused by pathogens. Innate defences include anatomical and structural barriers, inflammation, phagocytosis and the presence of a normal bacterial flora. The innate defences have also been referred to as "natural" or "consitutive" resistance, since they are inherent to the host.
2. *Inducible Defences*: Defence mechanisms that must be induced or turned on by host exposure to a pathogen (as during an infection). Unlike the innate defences, they are not immediately ready to come into play until after the host is appropriately exposed to the parasite. The inducible defences are synonymous with acquired or adaptive immunity and involve the immunological responses to a pathogen causing an infection.

Adaptive immunity is generally quite specifically directed against an invading pathogen. The innate defences are not so specific, and are directed towards general strategic defence. Innate defences, by themselves, may not be sufficient to protect a host against pathogens. Such pathogens that evade or overcome the relatively non-specific innate defences are usually susceptible to the more specific inducible defences, once they have developed.

Special note. Most immunologists have subverted some of the "innate" defences and moved them to the "inducible" category, although these defences are not usually thought of as part of the immunological system. This refers to complement activation, the inflammatory response and the phagocytic response. Their reasoning is that these responses are, in fact, elicited or turned on by some chemical, physical or biological stimulation. However, the components or cells involved are constitutive components of the host. Nonetheless, these innate responses to pathogens may initiate, participate with, or otherwise affect an immunological response.

THE IMMUNE SYSTEM OF ANIMALS

The inducible defences are so-called because they are induced upon primary exposure to a pathogen or one of its products. The inducible defences are a function of the immunological system and the immune responses. The innate defences and immediately available for host defence. The inducible defences must be triggered in a host and initially take time to develop. The type of resistance thus developed in the host is called acquired immunity. The term immune usually means the ability to resist infectious disease. Immunity refers to the relative state of resistance of the host to a specific pathogen brought on by the activities of the immunological system. Acquired or Adaptive Immunity, itself, is sometimes divided into two types, based on how it is acquired by the host. In active immunity, the host undergoes an immunological response and produces the cells and factors responsible for the immunity, *i.e.*, the host produces its own antibodies and/or immunoreactive lymphocytes. Active immunity can persist a long time in the host, up to many years in humans.

In passive immunity there is acquisition by a host of immune factors which were produced in another animal, *i.e.*, the host receives antibodies and/or immunoreactive lymphocytes originally produced in another animal. Passive immunity is typically short-lived and usually persists only a few weeks or months.

ANTIGENS

Antigens are chemical substances of relatively high molecular weight, that stimulate the immune response in animals. Bacteria are composed of various macromolecular components that are antigens or " antigenic" in their host and bacterial antigens interact with the host immunological system in a variety of ways.

NATURAL ANTIBODIES

Studies on germ-free animals have confirmed that a normal bacterial flora in the gastrointestinal tract are necessary for full development of immunological (lymphatic) tissues in the intestine. Furthermore, the interaction between these immune tissues and intestinal bacteria results in the production of serum and secretory antibodies that are directed against bacterial antigens. These antibodies probably help protect the host from invasion by its own normal flora, and they can cross react with antgenically-related pathogens. For example, antibodies against normal *E. coli* could react with closely-related pathogenic *Shigella dysenteriae*. These type of antibodies are sometimes called natural or cross-reactive antibodies.

BACTERIAL ANTIGENS MADE INTO VACCINES

In another way, bacterial antigens that are the components or products of pathogens are the substances that induce the immune defences of the host to defend against, and to eliminate, the pathogen or disease. In the laboratory, these bacterial antigens can be manipulated or changed so that they will stimulate the immune response in the absence of infection or pathology. These isolated or modified antigens are the basis for active immunisation (vaccination) against bacterial disease. Thus, a modified form of the tetanus toxin (tetanus toxoid), which has lost its toxicity but retains its antigenicity, is used to immunise against tetanus. Or, antigenic parts of the whooping cough bacterium, *Bordetella pertussis,* can be used to induce active formation of antibodies that will react with the living organism and thereby prevent infection.

ANTIMICROBIAL AGENTS

One line of defence against bacterial infection is chemotherapy with antimicrobial agents such as antibiotics. The ecological relationships between animals and bacteria in the modern world are mediated by the omnipresence of antibiotics. Antibiotics are defined as substances produced by a microorganism that kill or inhibit other microorganisms. Originally, a group of soil bacteria, the *Streptomyces*, were the most innovative producers of antibiotics for clinical usage. They were the source of streptomycin, tetracycline, erythromycin and chloramphenicol, to name just a few antibiotics. Because bacteria evolve rapidly towards resistance, because bacteria can exchange genes for antibiotic resistance, and because we have overused and misused antibiotics, many pathogens are emerging as resistant to antibiotics. There have already been reported infections by *Enterococcus*, *Staphylococcus aureus* and *Pseudomonas aeruginosa* that are refractory to all known antibiotics. Bacterial resistance to antimicrobial agents has become part of a pathogen's determinants of virulence. These are examples of genetic means by which bacteria exert their virulence.

The usage of antibiotics to control the growth of parasites is an artificial way to intervene in the natural process of the host-parasite interaction. But, of course, it is done for the obvious purpose of curing the disease. The body heals itself: most antibiotics just stop bacterial growth, and the host must rely entirely on its native defences to accomplish the neutralisation of bacterial toxins or the elimination of bacterial cells. The judicious use of antibiotics in the past five decades has saved millions of lives from infections caused by bacteria.

PROKARYOTIC CELL ARRANGEMENTS

Binary fission can produce two type of cell arrangements. The plane in which they divide will have the daughter cells either separated or still attached to one another. Cocci that remain in pairs are called "Diplococci". Cocci in chains are called "Streptococci". Cocci that divide in 2 planes and stay attached are called "Sarcinae". Cocci in clusters like bunches of grapes are called "Staphylococcus". Bacillus or rod-shaped bacteria divide transversely, which is perpendicular to the long axis. Most separate and become single cells, while others form either chains or pairs. Corynebacterium diphtheriae, which causes diphtheria, are one of the gram positive bacteria that divide by snapping division. By that type of division, a "V-Form occurs as the cells remain attached to one another. If they create a side-by-side arrangement, they are called Palisade. Bacillus is a morphological term (rod shaped) and written as Bacillus, it is a genus. Sarcinae is a morphological term (cuboidal) and written as Sarcina, it is a genus.

ENDOSPORES

Endospores" can be formed by the two genus, Bacillus and Clostr-idium. They are possibly virulent in this form as well as very durable.

A normal, viable, living, metabolizing and replicating cell is called a ' vegetative cell. This cell only produces "one" endospore. The endospore can germinate back into a vegetative cell. An endospore is a bacterial form for preservation and these are not really reproductive structures. Endospores form because the environmental conditions are not favourable.

Sporulation – Endospore formation

The bacterial DNA is replicated. DNA aligns the long axis of the cell and is attached to the inside of the plasma membrane. There is invagination of the membrane and a Forespore is formed. The forespore is surrounded by growth of the cytoplasmic membrane within a second membrane. The vegetative cells DNA is now disintegrating. Between the two membranes, a cortex of Dipicolinic Acid and Calcium ions are deposited.

The endospore is now formed and is covered by a spore coat. Some endospores cause increased swelling within the original cell. These endospores are Centrally, Terminally or Subterminally formed. Endospores are very resistant, especially to chemicals, heat, desiccation and radiation. The endospore matures and spore coat completes is formation. There is produce now, an increase in resistance to chemicals and eat. Conditions get worse and the endospore "pops" out of the original cell. It is now called a "free spore". If environmental conditions improve, the spore will germinate back into a viable vegetative cell, with metabolism.

In 1735, Carlos Linnaeus formulated a system of nomenclature that is still in use today. Scientists used the Latin language to name things and all scientific terms were then labeled with a Latin name. Each scientific name consists of two terms or names. The Genus, which has it first letter capitalize and the other letters are in the small case.

The species, in which all letters are written in small case. A scientific name is "ALWAYS" either underlined or italicized! Scientific names are coined because they describe an organism, honour a researcher or they identify a species niche or habitat.

3

Types of Bacteria

BACILLUS OF BACTERIA

Bacillus is a genus of Gram-positive, rod-shaped bacteria and a member of the phylum Firmicutes. *Bacillus* species can be obligate aerobes or facultative anaerobes, and test positive for the enzyme catalase. Ubiquitous in nature, *Bacillus* includes both free-living and pathogenic species. Under stressful environmental conditions, the cells produce oval endospores that can stay dormant for extended periods. These characteristics originally defined the genus, but not all such species are closely related, and many have been moved to other genera.

INDUSTRIAL SIGNIFICANCE

Many *Bacillus* species are able to secrete large quantities of enzymes. *Bacillus amyloliquefaciens* is the source of a natural antibiotic protein barnase (a ribonuclease), alpha amylase used in starch hydrolysis, the protease subtilisin used with detergents, and the BamH1 restriction enzyme used in DNA research.

A portion of the *Bacillus thuringiensis* genome was incorporated into corn (and cotton) crops. The resulting GMOs are therefore resistant to some insect pests.

USE AS MODEL ORGANISM

Bacillus subtilis is one of the best understood prokaryotes, in terms of molecular biology and cell biology. Its superb genetic amenability and relatively large size have provided the powerful tools required to investigate a bacterium from all possible aspects.

Recent improvements in fluorescence microscopy techniques have provided novel and amazing insight into the dynamic structure of a single cell organism. Research on *Bacillus subtilis* has been at the forefront of bacterial molecular biology and cytology, and the organism is a model for differentiation, gene/protein regulation, and cell cycle events in bacteria.

CLINICAL SIGNIFICANCE

Two *Bacillus* species are considered medically significant: *B. anthracis*, which causes anthrax, and *B. cereus*, which causes a foodborne illness similar to that of *Staphylococcus*. A third species, *B. thuringiensis*, is an important insect pathogen, and is sometimes used to control insect pests. The type species is *B. subtilis*, an important model organism. It is also a notable food spoiler, causing ropiness in bread and related food. Some environmental and commercial strains *B. coagulans* may play a role in food spoilage of highly acidic, tomato based products.

An easy way to isolate *Bacillus* is by placing non-sterile soil in a test tube with water, shaking, placing in melted mannitol salt agar, and incubating at room temperature for at least a day. Colonies are usually large, spreading and irregularly shaped. Under the microscope, the *Bacillus* cells appear as rods, and a substantial portion usually contain an oval endospore at one end, making it bulge.

CELL WALL

The cell wall of *Bacillus* is a structure on the outside of the cell that forms the second barrier between the bacterium and the environment, and at the same time maintains the rod shape and withstands the pressure generated by the cell's turgor. The cell wall is composed of teichoic and teichuronic acids. *B. subtilis* is the first bacterium for which the role of an actin-like cytoskeleton in cell shape determination and peptidoglycan synthesis was identified, and for which the entire set of peptidoglycan-synthesizing enzymes was localised. The role of the cytoskeleton in shape generation and maintenance is important.

PHYLOGENY

The genus *Bacillus* was coined in 1835 by Christian Gottfried Ehrenberg (who coined the genus *Bacterium* seven years prior) to contain rod-shaped bacteria, later amended by Ferdinand Cohn to spore-forming, Gram-positive/variable, rod-shaped bacteria. Like other genera associated with the early history of microbiology, such as *Pseudomonas* or *Vibrio*, members of the *Bacillus* genus (266 species) are found ubiquitously, and it is one of the genera with the largest 16S diversity and environmental diversity.

Several studies have tried to reconstruct the phylogeny of the genus. The *Bacillus*-specific study with the most diversity covered is by Xu and Cote' using 16S and the ITS region, where they divide the genus into 10 groups, which includes the nested genera *Paenibacillus, Brevibacillus, Geobacillus, Marinibacillus* and *Virgibacillus*. However, the tree constructed by the living tree project, a collaboration between ARB-Silva and LPSN where a 16S (and 23S if available) tree of all validated species was constructed, the genus *Bacillus* contains a very large number of nested taxa and majorly in both 16S and 23S it is paraphyletic to Lactobacillales (*Lactobacillus, Streptococcus, Staphylococcus, Listeria*, etc.), due to *Bacillus coahuilensis* and others.

A gene concatenation study found similar results to Xu and Cote', but with a much more limited number of species in terms of groups, but used *Listeria* as an outgroup, so in light of the ARB tree, it may be "inside-out".

PAENIBACILLUSOF BACTERIA

Paenibacillus is a genus of facultative anaerobic, endospore-forming bacteria, originally included within the genus *Bacillus* and then reclassified as a separate genus in 1993. Bacteria belonging to this genus have been detected in a variety of environments such as: soil, water, rhizosphere, vegetable matter, forage and insect larvae, as well as clinical samples.

The name reflects this fact: Latin *paene* means almost, and so the *Paenibacilli* are literally *almost Bacilli*. The genus includes *P. larvae*, which is known to cause American1d in agriculture and horticulture, the *Paenibacillus* sp. JDR-2 which is known to be a rich source of chemical agents for biotechnology applications and pattern forming strains such as *P. vortex* and *P. dendritiformis* discovered in the early 90s, which are known to develop complex colonies with intricate architectures as is illustrated in the pictures.

IMPORTANCE

There has been a rapidly growing interest in *Paenibacillus* spp. since many were shown to be important for agriculture and horticulture (*e.g. P. polymyxa*), industrial (*e.g. P. amylolyticus*), and medical applications (*e.g. P. peoriate*). These bacteria produce various extracellular enzymes such as polysaccharide-degrading enzymes and proteases, which can catalyse a wide variety of synthetic reactions in fields ranging from cosmetics to biofuel production. Various *Paenibacillus spp.* also produce antimicrobial substances that affect a wide spectrum of micro-organisms such as fungi, soil bacteria, plant pathogenic bacteria and even important anaerobic pathogens as *Clostridium botulinium*.

More specifically, several Paenibacillus species serve as efficient plant growth promoting rhizobacteria (PGPR). PGPR competitively colonise plant roots and can simultaneously act as biofertilizers and as antagonists (biopesticides) of recognised root pathogens, such as bacteria, fungi and nematodes. They enhance plant growth by several direct and indirect mechanisms.

Direct mechanisms include phosphate solubilisation, nitrogen fixation, degradation of environmental pollutants and hormone production. Indirect mechanisms include controlling phytopathogens by competing for resources such as iron, amino acids and sugars, as well as by producing antibiotics or lytic enzymes. Competition for iron also serves as a strong selective force determining the microbial population in the rhizosphere. Several studies show that PGPR exert their plant growth-promoting activity by depriving native microflora of iron. Although iron is abundant in nature, the extremely low solubility of Fe^{3+} at pH 7 means that most organisms face the problem of obtaining enough iron from their environment.

To fulfil their requirements for iron, bacteria have developed several strategies, including:

- The reduction of ferric to ferrous ions,
- The secretion of high-affinity iron-chelating compounds, called siderophores, and
- The uptake of heterologous siderophores.

P. vortex's genome for example, harbours many genes which are employed in these strategies, in particular it has the potential to produce siderophores under iron limiting conditions.

Despite the increasing interest in *Paenibacillus spp.* genomic information of these bacteria is lacking. More extensive genome sequencing could provide fundamental insights into pathways involved in complex social behaviour of bacteria, and can discover a rich source of genes with biotechnological potential.

PATTERN FORMATION, SELF-ORGANISATION AND SOCIAL BEHAVIOURS

Several *Paenibacillus* species can form complex patterns on semi-solid surfaces. Development of such complex colonies require self-organisation and cooperative behaviour of individual cells while employing sophisticated chemical communication. Pattern formation and self-organisation in microbial systems is an intriguing phenomenon and reflects social behaviours of bacteria that might provide insights into the evolutionary development of the collective action of cells in higher organisms.

Pattern Forming in *P. vortex*

One of the most fascinating pattern forming *Paenibacillus* species is *P. vortex*, self-lubricating, flagella driven bacteria. *P. vortex* organises its colonies by generating modules, each consisting of many bacteria, which are used as building blocks for the colony as a whole. The modules are groups of bacteria that move around a common center at about 10 μm/s.

Pattern Forming in *P. dendritiformis*

An additional intriguing pattern forming *Paenibacillus* species is *P. dendritiformis*, which is known to be able to generate two different morphotypes – the Branching (or tip-splitting) morphotype and the Chiral morphotype that is marked by curly branches with well defined handedness.

These two pattern forming *Paenibacillus* strains exhibit many distinct physiological and genetic traits including β-galactosidase-like activity causing colonies to turn blue on X-gal plates and multiple drug resistance (MDR) (including septrin, penicillin, kanamycin, chloramphenicol, ampicillin, tetracycline, spectinomycin, streptomycin and mitomycin C. Colonies that are grown on surfaces in Petri dishes exhibit several folds higher drug resistance in comparison to growth in liquid media. This particular resistance is believed to be due to a surfactant-like liquid front that actually forms a particular pattern on the Petri plate.

CLOSTRIDIUM OF BACTERIA

Clostridium is a genus of Gram-positive bacteria, belonging to the Firmicutes. They are obligate anaerobes capable of producing endospores. Individual cells are rod-shaped, which gives them their name, from the Greek kloster (klwsthr) or spindle. These characteristics traditionally defined the genus; however many species originally classified as Clostridium have been reclassified in other genera.

OVERVIEW

Clostridium consists of around 100 species that include common free-living bacteria as well as important pathogens.

There are five main species responsible for disease in humans:

1. *C. botulinum*, an organism that produces botulinum toxin in food/ wound and can cause botulism. Honey sometimes contains spores of *Clostridium botulinum*, which may cause infant botulism in humans one year old and younger. The toxin eventually paralyses the infant's breathing muscles. Adults and older children can eat honey safely, because *Clostridium* do not compete well with the other rapidly growing bacteria present in the gastrointestinal tract. This same toxin is known as "Botox" and is used cosmetically to paralyse facial muscles to reduce the signs of aging; it also has numerous therapeutic uses.
2. *C. difficile*, which can flourish when other bacteria in the gut are killed during antibiotic therapy, leading to pseudomembranous colitis (a cause of antibiotic-associated diarrhea).
3. *C. perfringens*, formerly called *C. welchii*, causes a wide range of symptoms, from food poisoning to gas gangrene. Also responsible for enterotoxemia (also known as "overeating disease" or "pulpy kidney disease") in sheep and goats. *C. perfringens* also takes the place of yeast in the making of salt rising bread. The name perfringens means 'breaking through' or 'breaking in pieces'.
4. *C. tetani*, the causative organism of tetanus. The name derives from the Greek word τετανoζ (tetanos) "muscular spasm", due to the violent spasms caused by C. tetani infection.

5. *C. sordellii* can cause a fatal infection in exceptionally rare cases after medical abortions. Less than one case per year has been reported since 2000.

Clostridium is sometimes found in raw swiftlet nests, a Chinese delicacy. Nests are washed in a sulphite solution to kill the bacteria before being exported to the U.S.

Neurotoxin production is the unifying feature of the species *C. botulinum*. Seven types of toxins have been identified and allocated a letter (A-G). Most strains produce one type of neurotoxin but strains producing multiple toxins have been described. *C. botulinum* producing B and F toxin types have been isolated from human botulism cases in New Mexico and California. The toxin type has been designated Bf as the type B toxin was found in excess of the type F. Similarly, strains producing Ab and Af toxins have been reported. Organisms genetically identified as other *Clostridium* species have caused human botulism; *Clostridium butyricum* producing type E toxin and *Clostridium baratii* producing type F toxin. The ability of *C. botulinum* to naturally transfer neurotoxin genes to other *Clostridium* species is concerning, especially in the food industry where preservation systems are designed to destroy or inhibit only *C. botulinum* but not other *Clostridium* species.

COMMERCIAL USES

C. thermocellum can utilise lignocellulosic waste and generate ethanol, thus making it a possible candidate for use in production of ethanol fuel. It also has no oxygen requirement and is thermophilic, which reduces cooling cost.

C. acetobutylicum, also known as the *Weizmann organism*, was first used by Chaim Weizmann to produce acetone and biobutanol from starch in 1916 for the production of gunpowder and TNT.

C. botulinum produces a potentially lethal neurotoxin that is used in a diluted form in the drug *Botox*, which is carefully injected to nerves in the face, which prevents the movement of the expressive muscles of the forehead, to delay the wrinkling effect of ageing. It is also used to treat spasmodic torticollis and provides relief for approximately 12 to 16 weeks.

The anaerobic bacterium *C. ljungdahlii*, recently discovered in commercial chicken wastes, can produce ethanol from single-carbon sources including synthesis gas, a mixture of carbon monoxide and hydrogen that can be generated from the partial combustion of either fossil fuels or biomass. Use of these bacteria to produce ethanol from synthesis gas has progressed to the pilot plant stage at the BRI Energy facility in Fayetteville, Arkansas.

Fatty acids are converted by yeasts to long-chain dicarboxylic acids and then to 1,3-propanediol using *Clostridium diolis*.

Genes from *C. thermocellum* have been inserted into transgenic mice to allow the production of endoglucanase. The experiment was intended to learn more about how the digestive capacity of monogastric animals could be improved. Hall *et al.* published their findings in 1993.

Non-pathogenic strains of *Clostridium* may help in the treatment of diseases such as cancer. Research shows that *Clostridium* can selectively target cancer cells. Some strains can enter and replicate within solid tumours. *Clostridium* could, therefore, be used to deliver therapeutic proteins to tumours. This use of *Clostridium* has been demonstrated in a variety of preclinical models.

Mixtures of *Clostridium* species, such as a mixtures of *C. beijerinckii*, *C. butyricum*, and species from other genera have been shown to produce biohydrogen from yeast waste.

CLOSTRIDIUM PERFRINGENS

Clostridium perfringens (formerly known as *C. welchii*) is a Gram-positive, rod-shaped, anaerobic, spore-forming bacterium of the genus *Clostridium*. *C. perfringens* is ever present in nature and can be found as a normal component of decaying vegetation, marine sediment, the intestinal tract of humans and other vertebrates, insects, and soil.

C. perfringens is the third most common cause of food poisoning in the United Kingdom and the United States though it can sometimes be ingested and cause no harm.

Infections due to *C. perfringens* show evidence of tissue necrosis, bacteremia, emphysematous cholecystitis, and gas gangrene, which is also known as clostridial myonecrosis. The toxin involved in gas gangrene is known as a-toxin, which inserts into the plasma membrane of cells, producing gaps in the membrane that disrupt normal cellular function. *C. perfringens* can participate in polymicrobial anaerobic infections. *Clostridium perfringens* is commonly encountered in infections as a component of the normal flora. In this case, its role in disease is minor.

The action of *C. perfringens* on dead bodies is known to mortuary workers as tissue gas and can be halted only by embalming.

FOOD POISONING

In the United Kingdom and United States, *C. perfringens* bacteria are the third most common cause of foodborne illness, with poorly prepared meat and poultry, or food properly prepared but left to stand too long, the main culprits in harbouring the bacterium. The clostridium perfringens enterotoxin (CPE) mediating the disease is heat-labile (inactivated at 74 °C (165 °F)) and can be detected in contaminated food, if not heated properly, and feces. Incubation time is between six and 24 (commonly 10-12) hours after ingestion of contaminated food. Since *C. perfringens* forms spores that can withstand cooking temperatures, if cooked food is let stand for long enough, germination can ensue and infective bacterial colonies develop. Symptoms typically include abdominal cramping, diarrhea; vomiting and fever are usual. The whole course usually resolves within 24 hours. Very rare, fatal cases of clostridial necrotising enteritis (also known as pigbel) have been known to involve "Type C" strains of the organism, which produce a potently ulcerative β-toxin. This strain is most frequently encountered in Papua New Guinea.

Many cases of *C. perfringens* food poisoning likely remain subclinical, as antibodies to the toxin are common among the population. This has led to the conclusion that most of the population has experienced food poisoning due to *C. perfringens*.

Despite its potential dangers, *C. perfringens* is used as the leavening agent in salt rising bread. The baking process is thought to reduce the bacterial contamination, precluding negative effects.

INFECTION

Clostridium perfringens is the most common bacterial agent for gas gangrene, which is necrosis, putrefaction of tissues, and gas production. It is caused primarily by Clostridium perfringens alpha toxin. The gases form bubbles in muscle (crepitus) and the characteristic smell in decomposing tissue. After rapid and destructive local spread (which can take only hours), systemic spread of bacteria and bacterial toxins may cause death. This is a problem in major trauma and in military contexts. *C. perfringens* grows readily on blood agar plate in anaerobic conditions, and often produces a double zone of beta hemolysis.

DIAGNOSIS

C. perfringens can be diagnosed by Nagler's Reaction where the suspect organism is cultured on an egg yolk media plate. One side of the plate contains anti-alpha-toxin, while the other side does not. A streak of suspect organism is placed through both sides. An area of turbidity will form around the side that does not have the anti-alpha-toxin, indicating uninhibited lecithinase activity.

TREATMENT

If detected on clinical ground, treatment should begin without waiting for lab results. Traumatic wounds should be cleaned. Wounds that cannot be cleaned should not be stitched shut. Penicillin prophylaxis kills many clostridia, and is thus useful for dirty wounds and lower leg amputations. A high infectious dose is required; the carrier state persists for several days.

BRUCELLA CAUSES

Brucella is a genus of Gram-negative bacteria named after David Bruce (1855-1931). They are small (0.5 to 0.7 by 0.6 to 1.5 μm), non-motile, non-encapsulated coccobacilli, which function as facultative intracellular parasites.

Brucella is the cause of brucellosis, which is a zoonosis. It is transmitted by ingesting contaminated food (such as unpasteurised milk products), direct contact with an infected animal, or inhalation of aerosols. Transmission from human to human, for example through sexual intercourse or from mother to child, is exceedingly rare, but possible. Minimum infectious exposure is between 10 - 100 organisms. The different species of *Brucella* are genetically very similar although each has a slightly different host specificity. Hence the NCBI taxonomy includes most *Brucella* species under *Brucella melitensis*.

HUMAN BRUCELLOSIS

Sir David Bruce isolated *B. melitensis* from British soldiers who died from Malta fever in Malta. After exposure to *Brucella*, humans generally have a two- to four-week latency period before exhibiting symptoms. Symptoms include acute undulating fever (>90 per cent of all cases), headache, arthralgia (>50 per cent), night sweats, fatigue and anorexia. Later complications may include arthritis or epididymoorchitis, spondylitis, neurobrucellosis, liver abscess formation, and endocarditis, the latter potentially fatal.

Human brucellosis is usually not transmitted from human to human; people become infected by contact with fluids from infected animals (sheep, cattle or pigs) or derived food products like unpasteurised milk and cheese. Brucellosis is also considered an occupational disease because of a higher incidence in people working with animals (slaughterhouse cases). People may also be infected by inhalation of contaminated dust or aerosols. Human and animal brucellosis share the persistence of the bacteria in tissues of the mononuclear phagocyte system, including the spleen, liver, lymph nodes, and bone marrow. *Brucella* can also target the male reproductive tract. Globally, there are an estimated 500,000 cases of brucellosis each year.

DIAGNOSIS

Brucella is isolated from a blood culture on Castaneda medium. Prolonged incubation (up to 6 weeks) may be required as they are slow-growing, but on modern automated machines the cultures often show positive results within seven days. On Gram stain they appear as dense clumps of Gram-negative coccobacilli and are exceedingly difficult to see. In recent years molecular diagnostic techniques based on the genetic component of the pathogen have become more popular. It is crucial to be able to differentiate *Brucella* from *Salmonella* which could also be isolated from blood cultures and are Gram-negative. Testing for urease would successfully accomplish the task; it is positive for *Brucella* and negative for *Salmonella*. Brucella can also be seen in bone marrow biopsies. Laboratory acquired brucellosis is common. This most often happens when the disease is not thought of until cultures become positive, by which time the specimens have already been handled by a number of laboratory staff. The idea of preventive treatment is to stop people who have been exposed to *Brucella* from becoming ill with the disease.

TREATMENT

There are no clinical trials to be relied on as a guide for optimal treatment, but a three week course of rifampicin and doxycycline twice daily is the combination most often used, and appears to be efficacious; the advantage of this regimen is that it is oral medication and there are no injections; however, a high rate of side effects (nausea, vomiting, loss of appetite) has also been reported.

HOST SPECIFICITY AND ANIMAL BRUCELLOSIS

Pathogenic *Brucella* can cause abortion in female animals by colonisation of placental trophoblasts, and sterility in male animal.

POSSIBLE PATHOGENS RESPONSIBLE FOR THE PLAGUE IN THEBES

According to a January 2012 article in *Emerging Infectious Diseases*, pathogens that fit the characteristics of the plague described in *Oedipus Rex* include Leishmania spp., Leptospira spp., Brucella abortus, Orthopoxviridae, and Francisella tularensis. While the authors note that the disease progression of brucellosis in modern times may make it seem unlikely, they posit that it may have been an agent impacting the livestock, while humans might have been infected with Salmonella enterica serovar Typhi or another pathogen or that the ancestral versions of Brucella may have been more lethal.

BRUCELLOSIS

Brucellosis, also called Bang's disease, Crimean fever, Gibraltar fever, Malta fever, Maltese fever, Mediterranean fever, rock fever, or undulant fever, is a highly contagious zoonosis caused by ingestion of unsterilised milk or meat from infected animals or close contact with their secretions. Transmission from human to human, through sexual contact or from mother to child, is rare but possible.

Brucella spp. are small, Gram-negative, non-motile, non-spore-forming, rod shaped (coccobacilli) bacteria. They function as facultative intracellular parasites causing chronic disease, which usually persists for life. Symptoms include profuse sweating and joint and muscle pain. Brucellosis has been recognised in animals including humans since the 20th century.

SIGNS AND SYMPTOMS

The symptoms are like those associated with many other febrile diseases, but with emphasis on muscular pain and sweating. The duration of the disease can vary from a few weeks to many months or even years. In the first stage of the disease, septicaemia occurs and leads to the classic triad of undulant fevers, sweating (often with characteristic smell, likened to wet hay), and migratory arthralgia and myalgia. Blood tests characteristically reveal leukopenia and anemia, show some elevation of AST and ALT, and demonstrate positive Bengal Rose and Huddleston reactions.

This complex is, at least in Portugal, known as Malta fever. During episodes of Malta fever, melitococcemia (presence of brucellae in blood) can usually be demonstrated by means of blood culture in tryptose medium or Albini medium.

If untreated, the disease can give origin to focalisations or become chronic. The focalisations of brucellosis occur usually in bones and joints and spondylodiscitis of lumbar spine accompanied by sacroiliitis is very characteristic of this disease. Orchitis is also frequent in men.

Diagnosis of brucellosis relies on:

1. Demonstration of the agent: blood cultures in tryptose broth, bone marrow cultures. The growth of brucellae is extremely slow (they can take until 2 months to grow) and the culture poses a risk to laboratory personnel due to high infectivity of brucellae.
2. Demonstration of antibodies against the agent either with the classic Huddleson, Wright and/or Bengal Rose reactions, either with ELISA or the 2-mercaptoethanol assay for IgM antibodies associated with chronic disease
3. Histologic evidence of granulomatous hepatitis (hepatic biopsy)
4. Radiologic alterations in infected vertebrae: the Pedro Pons sign (preferential erosion of antero-superior corner of lumbar vertebrae) and marked osteophytosis are suspicious of brucellic spondylitis.

The disease's sequelae are highly variable and may include granulomatous hepatitis, arthritis, spondylitis, anaemia, leukopenia, thrombocytopenia, meningitis, uveitis, optic neuritis, endocarditis, and various neurological disorders collectively known as neurobrucellosis.

CAUSE

Brucellosis in humans is usually associated with the consumption of unpasteurised milk and soft cheeses made from the milk of infected animals, primarily goats, infected with *Brucella melitensis* and with occupational exposure of laboratory workers, veterinarians, and slaughterhouse workers. Some vaccines used in livestock, most notably *B. abortus* strain 19, also cause disease in humans if accidentally injected. Brucellosis induces inconstant fevers, abortion, sweating, weakness, anaemia, headaches, depression, and muscular and bodily pain.

TREATMENT AND PREVENTION

Antibiotics like tetracyclines, rifampicin, and the aminoglycosides streptomycin and gentamicin are effective against *Brucella* bacteria. However, the use of more than one antibiotic is needed for several weeks, because the bacteria incubate within cells. The gold standard treatment for adults is daily intramuscular injections of streptomycin 1 g for 14 days and oral doxycycline 100 mg twice daily for 45 days (concurrently). Gentamicin 5 mg/kg by intramuscular injection once daily for seven days is an acceptable substitute when streptomycin is not available or difficult to obtain. Another widely used regimen is doxycycline plus rifampin twice daily for at least six weeks. This regimen has the advantage of oral administration. A triple therapy of doxycycline, with rifampin and cotrimoxazole, has been used successfully to treat neurobrucellosis.

Doxycycline is able to cross the blood–brain barrier, but requires the addition of two other drugs to prevent relapse. Ciprofloxacin and cotrimoxazole therapy is associated with an unacceptably high rate of relapse. In brucellic endocarditis, surgery is required for an optimal outcome. Even with optimal antibrucellic therapy, relapses still occur in 5–10 percent of patients with Malta fever.

The main way of preventing brucellosis is by using fastidious hygiene in producing raw milk products, or by pasteurising all milk that is to be ingested by human beings, either in its unaltered form or as a derivate, such as cheese. Experiments have shown that cotrimoxazol and rifampin are both safe drugs to use in treatment of pregnant women who have brucellosis.

PROGNOSIS

With combination drug therapy, most individuals recover in 2 to 3 weeks. Even widespread infections may be cured. Untreated, however, the infection may progress and increase in severity and also affect new tissues. Although brucellosis can take a chronic form, with periods of illness alternating with periods of no symptoms, persistent illness lasting longer than 2 months may be due to a previously unsuspected underlying disease or a complication of the brucellosis.

Approximately 10 per cent of individuals may have a relapse, even after treatment is completed. In these cases, treatment should be repeated.

This disease has a low mortality rate (lower than 2 per cent); the most likely cause of death is endocarditis caused by Brucella melitensis.

BIOLOGICAL WARFARE

In 1954, *B. suis* became the first agent weaponised by the United States at its Pine Bluff Arsenal in Arkansas. *Brucella* species survive well in aerosols and resist drying. *Brucella* and all other remaining biological weapons in the U.S., arsenal were destroyed in 1971–72 when the U.S., offensive biological weapons (BW) programme was discontinued.

The United States BW programme focused on three agents of the Brucella group:

1. Porcine Brucellosis (Agent US)
2. Bovine Brucellosis (Agent AB)
3. Caprine Brucellosis (Agent AM)

Agent US was in advanced development by the end of World War II. When the U.S. Air Force (USAF) wanted a biological warfare capability, the Chemical Corps offered Agent US in the M114 bomblet, based after the 4-pound bursting bomblet developed for anthrax in World War II. Though the capability was developed, operational testing indicated that the weapon was less than desirable, and the USAF termed it an interim capability until replaced by a more effective biological weapon.

The main drawbacks of the M114 with Agent US was that it was incapacitating (the USAF wanted "killer" agents), the storage stability was too low to allow for storing at forward air bases, and the logistical requirements to neutralise a target were far higher than originally anticipated, requiring unreasonable logistical air support.

Agents US and AB had a median infective dose of 500 org/person, and AM was 300 org/person. The rate-of-action was believed to be 2 weeks, with a duration of action of several months. The lethality estimate was based on epidemiological information at 1–2 per cent. AM was always believed to be a more virulent disease, and a 3 per cent fatality rate was expected.

HISTORY

Under the name *Malta fever*, the disease now called brucellosis first came to the attention of British medical officers in the 1850s in Malta during the Crimean War. The causal relationship between organism and disease was first established in 1887 by Dr. David Bruce. In 1897, Danish veterinarian Bernhard Bang isolated *Brucella abortus* as the agent; and the additional name Bang's disease was assigned.

Maltese doctor and archaeologist Sir Themistocles Zammit earned a knighthood for identifyingunpasteurised milk as the major source of the pathogen in 1905, and it has since become known as Malta Fever. In cattle, this disease is also known as contagious abortion and infectious abortion. The popular name undulant fever originates from the characteristic undulance (or "wave-like" nature) of the fever, which rises and falls over weeks in untreated patients. In the 20th century, this name, along with *brucellosis* (after *Brucella*, named for Dr. Bruce), gradually replaced the 19th century names Mediterranean fever and *Malta fever*.

In 1989, neurologists in Saudi Arabia discovered neurobrucellosis, a neurological involvement in brucellosis.

The following obsolete names have previously been applied to brucellosis:

- *Brucelliasis*
- Bruce's septicemia

- Chumble fever
- Continued fever
- Crimean fever
- Cyprus fever
- *Febris melitensis*
- *Febris undulans*
- Fist of mercy
- Goat fever
- *Melitensis septicemia*
- Melitococcosis
- Milk sickness
- Mountain fever
- Neapolitan fever
- Satan's fever
- Slow fever
- Scottish Delight
- Jones Disease
- Contagious abortion

BRUCELLOSIS IN DOGS

The causative agent of brucellosis in dogs is *Brucella canis*. It is transmitted to other dogs through breeding and contact with aborted fetuses. Brucellosis can occur in humans that come in contact with infected aborted tissue or semen. The bacteria in dogs normally infect the genitals and lymphatic system, but can also spread to the eye, kidney, and intervertebral disc (causingdiscospondylitis). Symptoms of brucellosis in dogs include abortion in female dogs and scrotal inflammation and orchitis(inflammation of the testicles) in males. Fever is uncommon. Infection of the eye can cause uveitis, and infection of the intervertebral disc can cause pain or weakness. Blood testing of the dogs prior to breeding can prevent the spread of this disease. It is treated with antibiotics, as with humans, but it is difficult to cure.

BRUCELLA CANISOF BACTERIA

Brucella canis is a gram-negative proteobacterium in the family Brucellaceae that causes brucellosis in dogs and other canids. Bacteria *B. canis* are rod-shaped or cocci, oxidase, catalase, and urease positive. The species was firstly described in USA in 1966 where mass abortions of beagles were documented. The disease is characterised by epididymitis and orchitis in male dogs, endometritis, placentitis and abortions in females, and often presents as infertility in both sexes. Other symptoms such as inflammation in eyes and axial and appendicular skeleton, lymphadenopathy and splenomegaly, are less common. Humans can be also infected but infections are rare. *Brucella canis* is a zoonotic organism. Signs of this disease are different in both genders of dogs; females who have *Brucella canis* face an abortion of their pre-developed fetus. Males face the chance of infertility; the reason is that they develop an antibody against the sperm. This may be followed by inflammation of the testes which will generally settle down a little while after. Symptoms do not only include testicular inflammation, infertility in males, and abortion in females. Another symptom is the infection of the spinal plates or vertebrae, which is called diskospondylitis.

Treatment for *B. canis* is very difficult to find and often very expensive. The combination of minocycline and streptomycin is *thought* to be useful, but it is often unaffordable. Tetracycline can be a less expensive substitute for minocycline, but it also lowers the effect of the treatment.

BURKHOLDERIA PSEUDOMALLEI

Burkholderia pseudomallei (also known as *Pseudomonas pseudomallei*) is a Gram-negative, bipolar, aerobic, motile rod-shaped bacterium. It infects humans and animals and causes the disease melioidosis. It is also capable of infecting plants.

B. pseudomallei measures 2-5 ìm in length and 0.4-0.8 ìm in diameter and are capable of self-propulsion using flagellae. The bacteria can grow in a number of artificial nutrient environments, especially betaine- and arginine-containing. *in vitro*, optimal proliferation temperature is reported around 40°C in pH-neutral or slightly acidic environments (pH 6.8–7.0). The majority of strains are capable of fermentation of sugars without gas formation (most importantly, glucose and galactose, older cultures are reported to also metabolise maltose and starch). Bacteria produce both exo- and endo-toxins. The role of the toxins identified in the process of melioidosis symptom development has not been fully elucidated.

IDENTIFICATION

B. pseudomallei is not fastidious and will grow on a large variety of culture media (blood agar, MacConkey agar, EMB, etc.). Ashdown's medium (or *Burkholderia cepacia* medium) may be used for selective isolation. Cultures typically become positive in 24 to 48 hours (this rapid growth rate differentiates the organism from *B. mallei*, which typically takes a minimum of 72 hours to grow). Colonies are wrinkled, have a metallic appearance, and possess an earthy odour. On Gram staining, the organism is a Gram-negative rod with a characteristic "safety pin" appearance (bipolar staining). On sensitivity testing, the organism appears highly resistant (it is innately resistant to a large number of antibiotics including colistin and gentamicin) and that again differentiates it from *B. mallei*, which is in contrast, exquisitely sensitive to a large number of antibiotics. For environmental specimens only, differentiation from the non-pathogenic *B. thailandensis* using an arabinose test is necessary (*B. thailandensis* is never isolated from clinical specimens). The laboratory identification of *B. pseudomallei* has been described in the literature.

The classic textbook description of *B. pseudomallei* in clinical samples is of an intracellular bipolar-staining Gram-negative rod, but this is of little value in identifying the organism from clinical samples. It has been suggested by some that the Wayson stain is useful for this purpose, but this has been shown not to be the case.

Laboratory identification of *B. pseudomallei* can be difficult, especially in Western countries where *B. pseudomallei* is rarely seen. The large wrinkled colonies look like environmental contaminants and are therefore often discarded as being of no clinical significance. Colony morphology is very variable and a single strain may display up multiple colony types, so inexperienced laboratory staff may mistakenly believe the growth is not pure. The organism grows more slowly than other bacteria that may be present in clinical specimens, and in specimens from non-sterile sites, is easily overgrown. Non-sterile specimens should therefore be cultured in selective media (*e.g.*, Ashdown's or B. cepacia medium. Even when the isolate is recognised to be significant, commonly used identification systems may misidentify the organism as *Chromobacterium violaceum* or other non-fermenting gram-negative bacilli such as *Burkholderia cepacia* or *Pseudomonas aeruginosa*. Again, because the disease is rarely seen in western countries, identification of the bacterium *B. pseudomallei* in cultures may not actually trigger alarm bells in physicians unfamiliar with the disease. Routine biochemical methods for identification of bacteria vary widely in their identification of this organism: the API 20NE system accurately identifies *B. pseudomallei* in 99 per cent of cases, as does the automated Vitek 1 system, but the automated Vitek 2 system only identifies 19 per cent of isolates.

The pattern of resistance to antimicrobials is distinctive, and helps to differentiate the organism from *P. aeruginosa*. The majority of *B. pseudomallei* isolates are intrinsically resistant to all aminoglycosides (via an efflux pump mechanism), but sensitive to co-amoxiclav: this pattern of resistance almost never occurs in *P. aeruginosa* and is helpful in identification.

Molecular methods (PCR) of diagnosis are possible, but not routinely available for clinical diagnosis. Fluorescence *in situ* hybridisation has also been described, but has not been clinical validated and it not commercially available. In Thailand, a latex agglutination assay is widely used. A rapid immunofluorescence technique is also available in a small number of centres in Thailand.

DISINFECTION

B. pseudomallei is susceptible to numerous disinfectants including benzalkonium chloride, iodine, mercuric chloride, potassium permanganate, 1 per cent sodium hypochlorite, 70 per cent ethanol, 2 per cent glutaraldehyde and to a lesser extent, phenolic preparations. *B. pseudomallei* is effectively killed by the commercial disinfectants, Perasafe and Virkon. The microorganism can also be destroyed by heating to above 74°C for 10 min or by UV irradiation. *B. pseudomallei* is not reliably disinfected by chlorine.

MEDICAL IMPORTANCE

B. pseudomallei infection in humans is called melioidosis. The mortality of melioidosis is 20 to 50 per cent even with treatment.

ANTIBIOTIC TREATMENT AND SENSITIVITY TESTING

The antibiotic of choice is ceftazidime. While various antibiotics are active *in vitro* (*e.g.*, chloramphenicol, doxycycline, co-trimoxazole), they have been proven to be inferior *in vivo* for the treatment of acute melioidosis. Disc diffusion tests are unreliable when looking for co-trimoxazole resistance in *B. pseudomallei* (they greatly overestimate resistance) and Etests or agar dilution tests should be used in preference. The actions of co-trimoxazole and doxycycline are antagonistic, which suggests that these two drugs ought not to be used together.

The organism is intrinsically resistant to gentamicin and to colistin, and this fact is helpful in the identification of the organism. Kanamycin is used to kill *B. pseudomallei* in the laboratory, but the concentrations used are much higher than those achievable in humans.

PATHOGENICITY MECHANISMS AND VIRULENCE FACTORS

B. pseudomallei is an "accidental pathogen". It is an environmental organism that has no requirement to pass through an animal host in order to replicate. From the point of view of the bacterium, human infection is an evolutionary "dead end".

Strains which cause disease in humans differ from those causing disease in other animals by possessing certain genomic islands. It may have the ability to cause disease in humans because of DNA it has acquired from other microorganisms. The mutation rate is also high, and the organism continues to evolve even after infecting the host.

B. pseudomallei is able to invade cells (it is an intracellular pathogen). It is able to polymerise actin and to spread from cell to cell, causing cell fusion and the formation of multinucleate giant cells. The bacterium also expresses a toxin called lethal factor 1. *B. pseudomallei* is one of the first proteobacteria to be identified as containing an active Type 6 secretion system. it is also the only organism identified that contains up to six different type 6 secretion systems. *B. pseudomallei* is intrinsically resistant to a large number of antimicrobial agents. One important mechanism is that it is able to pump drugs out of the cell, and this mediates resistance to aminoglycosides (*AmrAB-OprA*), tetracyclines, fluoroquinolones and macrolides (*BpeAB-OprB*).

VACCINE CANDIDATES

There are no vaccine currently available, but a number of vaccines candidates have been suggested. Aspartate-β-semialdehyde dehydrogenase (*asd*) gene deletion mutants are auxotrophic for diaminopimelate (DAP) in rich medium and auxotrophic for DAP, lysine, methionine, and threonine in minimal medium. The Δ*asd* bacterium (bacterium with the *asd* gene removed) protects against inhalational melioidosis in mice.

BURKHOLDERIA CEPACIA COMPLEX

Burkholderia cepacia complex (BCC), or simply *Burkholderia cepacia* is a group of catalase-producing, lactose-non-fermenting, Gram-negative bacteria composed of at least 17 different species, including *B. cepacia*, *B. multivorans*, *B. cenocepacia*, *B. vietnamiensis*, *B. stabilis*, *B. ambifaria*, *B. dolosa*, *B. anthina*, and *B. pyrrocinia*. *B. cepacia* is an important human pathogen which most often causes pneumonia in immunocompromised individuals with underlying lung disease (such as cystic fibrosis or chronic granulomatous disease). It also attacks young onion and tobacco plants, as well as displaying a remarkable ability to digest oil.

PATHOGENESIS

BCC organisms are typically found in water and soil and can survive for prolonged periods in moist environments. Person-to-person spread has been documented; as a result, many hospitals, clinics, and camps have enacted strict isolation precautions for those infected with BCC. Infected individuals are often treated in a separate area from uninfected patients to limit spread, since BCC infection can lead to a rapid decline in lung function and result in death.

DIAGNOSIS

Diagnosis of BCC involves culturing the bacteria from clinical specimens, such as sputum or blood. BCC organisms are naturally resistant to many common antibiotics, including aminoglycosides and polymyxin B. and this fact is exploited in the identification of the organism. Oxidation-fermentation polymyxin-bacitracin-lactose (OFPBL) agar contains polymyxin (which kills most Gram-negative bacteria, including *Pseudomonas aeruginosa*) and bacitracin (which kills most Gram-positive bacteria and *Neisseria* species). It also contains lactose, and organisms such as BCC that do not ferment lactose turn the pH indicator yellow, which helps to distinguish it from other organisms that may grow on OFPBL agar, such as *Candida* species, *Pseudomonas fluorescens*, *Stenotrophomonas* species, and *Proteus* species.

INFECTION CONTROL

The bacterium is so hardy, it has been found to persist in betadine (a common topical antiseptic). Recently, a 0.2 per cent chlorhexidine mouthwash was also recalled, after it was found to be contaminated with *B. cepacia*. On 1-August-2012, the US FDA announced a recall of selected lots of benzalkonium chloride swabs and antiseptic wipes manufactured for Dukal by Jianerkang Medical Dressing Co. Matrixx Initiatives Issues Nationwide Voluntary Recall of One Lot of Zicam® Extreme Congestion Relief Due to Contamination With Burkholderia Cepacia

TREATMENT

Treatment typically includes multiple antibiotics and may include ceftazidime, doxycycline, piperacillin, meropenem, chloramphenicol and trimethoprim/sulfamethoxazole(co-trimoxazole). Although co-trimoxazole has been generally considered the drug of choice for *B. cepacia* infections, ceftazidime, doxycycline, piperacillin and meropenem are considered to be viable alternative options in cases where co-trimoxazole cannot be administered because of hypersensitivity reactions, intolerance or resistance. In April 2007, researchers from the Schulich School of Medicine and Dentistry at the University of Western Ontario, working with a group from Edinburgh, announced that they had discovered a potential method to kill the organism, involving disruption in the biosynthesis of an essential cell membrane sugar.

HISTORY

B.cepacia was discovered by Walter Burkholder in 1949 as the cause of onion skin rot, and first described as a human pathogen in the 1950s. In the 1980s, it was first recognised in individuals with cystic fibrosis, and outbreaks were associated with a 35 per cent death rate. *B. cepacia* has a large genome, containing twice the amount of genetic material as *E. coli*.

MORAXELLA DISEASE

Moraxella is a genus of Gram-negative bacteria in the Moraxellaceae family. It is named after the Swiss ophthalmologist Victor Morax. The organisms are short rods, coccobacilli or, as in the case of *Moraxella catarrhalis*, diplococci in morphology, with asaccharolytic, oxidase-positive and catalase-positive properties. *Moraxella catarrhalis* is the clinically most important species under this genus.

ROLES IN DISEASE

The organisms are commensals of mucosal surfaces and sometimes give rise to opportunistic infection:

- *Moraxella catarrhalis* usually resides in respiratory tract, but can gain access to the lower respiratory tract in patients with chronic chest disease or compromised host defences, thus causing tracheobronchitis and pneumonia. For example, it causes a significant proportion of lower respiratory tract infections in elderly patients with COPD and chronic bronchitis. It is also one of the notable causes of otitis media and sinusitis. It causes similar symptoms to *Haemophilus influenzae*, although it is much less virulent. Unlike *Neisseria meningitidis*, which is a morphologic cousin of *Moraxella catarrhalis*, it hardly ever causes bacteremia or meningitis.

- *Moraxella lacunata* is one of the causes of blepharoconjunctivitis in human.
- *Moraxella bovis* is the cause of Infectious bovine keratoconjunctivitis; known colloquially in the United Kingdom as 'New Forest Eye'. As a strict aerobe, *M. bovis* is confined to the cornea and conjunctiva, resulting in a progressive, non-self-limiting keratitis, ulceration and - ultimately - rupture of the cornea. The disease is relatively common, infecting cattle only. Treatment is via the use of either subconjunctival injection of a Tetracycline, or topical application of Cloxacillin, the former being more effective. The bacterium can be transmitted by flies, so fly control may be necessary on farms throughout the summer. Rupture of the eye is more serious, and requires immediate enucleation, though the procedure itself carries a good prognosis.

MORAXELLA BOVIS

Moraxella bovis is a Gram-negative, aerobic, oxidase-positive diplococcus that is implicated in infectious bovine keratoconjunctivitis, an eye disease of cattle, also colloquially known as Pink Eye.

MORAXELLA CATARRHALIS INFECTION

Moraxella catarrhalis causes ear and upper and lower respiratory infections. Previously classified as Micrococcus, then Neisseria, and also known as Branhamella catarrhalis, this organism is a frequent cause of otitis media in children, acute and chronic sinusitis at all ages, and lower respiratory infection in adults with chronic lung disease. It is the 2nd most common bacterial cause of COPD exacerbations after non-typeable Haemophilus influenzae. M. catarrhalis pneumonia resembles pneumococcal pneumonia. Although bacteremia is rare, half of patients die within 3 mo because of intercurrent diseases. The prevalence of M. catarrhalis colonisation depends on age. About 1 to 5 per cent of healthy adults have upper respiratory tract colonisation. Nasopharyngeal colonisation with M. catarrhalis is common throughout infancy, may be increased during winter months, and is a risk factor for acute otitis media; early colonisation is a risk factor for recurrent otitis media. Substantial regional differences in colonisation rates occur. Living conditions, hygiene, environmental factors (eg, household smoking), genetic characteristics of the populations, host factors, and other factors may contribute to these differences.

The organism appears to spread contiguously from its colonising position in the respiratory tract to the infection site.

There is no pathognomonic feature of M. catarrhalis otitis media, acute or chronic sinusitis, or pneumonia. In lower respiratory disease, patients experience increased cough, purulent sputum production, and increased dyspnea.

These gram-negative cocci resemble Neisseria sp but can be readily distinguished by routine biochemical tests after culture isolation from infected fluids or tissues.

All strains now produce β-lactamase. The organism is generally susceptible to β-lactam/β-lactamase inhibitors, sulfamethoxazole, tetracyclines, extended-spectrum oral cephalosporins, aminoglycosides, macrolides, and fluoroquinolones.

PINK-EYE IN BEEF CATTLE

Pink-eye, or infectious bovine keratoconjunctivitis, is an inflammatory bacterial infection of the eye that can cause permanent blindness in severe cases. It is a contagious disease and in Victoria occurs mainly in young cattle in summer and autumn.

Pinkeye can affect up to 80 per cent of a mob, with affected weaner calves losing up to 10 per cent of their body weight. Although rare, deaths may occur when both eyes are affected, due to starvation, thirst or misadventure.

CAUSE

Most cases of pinkeye are caused by the bacterium Moraxella bovis of which there are several strains. The bacteria produce a toxin which attacks the surface of the eye (cornea) and conjuctiva, causing inflammation and ulceration. Several factors influence infection of the eye by Moraxella bovis. Infective material discharged from the eyes of affected cattle can be spread to other animals by flies, or onto long grass grazed by the cattle. Sunlight and dust make the problem worse. There may also be other organisms, such as viruses and mycoplasmas, which cause eye damage and allow Moraxella bovis to become established.

OCCURRENCE

Pink-eye is mainly a disease of young cattle commonly occurring in their first summer. Severe outbreaks may occur in older cattle if they have never been exposed to the disease. After infection, cattle develop a temporary immunity which lasts up to a year. Exposure to the causative agents in following years gives further immunity, usually without eye changes being obvious.

Infection can occasionally persist in a few animals and these are a source of infection in the following summer. The infection rate increases to a peak about 3-4 weeks after the first cases appear, and then gradually decreases.

The prevalence of pink-eye in districts and on individual farms varies from year to year, depending on seasons and weather, the fly population and whether cattle are grasing long grass. On some farms there may be only occasional cases while on others 60-80 per cent of cattle may be affected in very severe outbreaks. In most years the average infection rate is 5-10 per cent.

SIGNS AND COURSE OF PINK-EYE

The first signs seen are:

- Copious watery eye discharge
- Aversion to sunlight
- Signs of irritation: for example, excessive blinking
- Reddening and swelling of the eyelids and the third eyelid.

The eye will then go cloudy in the middle and the cornea may ulcerate during the next two days. Many animals spontaneously recover at this stage. In a small number of untreated cases, ulceration may progress to abscess formation, with possible rupture of the cornea and permanent blindness. After recovery about 2 per cent of affected eyes have a residual white scar on the cornea. An even smaller percentage progress to abscess formation. Most animals are completely recovered 3-5 weeks after infection without treatment and usually only one eye is affected. Animals that are severely affected in both eyes may grow poorly. Milking cows may produce less milk. Cattle become more difficult to muster with blind animals in the herd. Accidental deaths may occur if affected animals fall into dams or are trapped in fences.

OTHER CAUSES OF EYE TROUBLE

Grass Seeds

Grass seeds are also troublesome during the time of the year when pink-eye usually occurs. Examination of the eyelids, including behind the third eyelid may reveal one or more grass seeds. Damage to the eye caused by a grass seed is generally more varied than damage due to pink-eye, and the eye does not get better unless the grass seed is removed.

Cancer of the Eye

Eye cancer (squamous cell carcinoma) usually originates on the unpigmented lateral part of the cornea (sclera) but some cancers will begin on unpigmented eyelids or the third eyelid. Initially most cancers begin as a plaque or growth with ragged edges on the lateral corneal surface. Some plaques will disappear, but many will grow into larger eye cancers if left untreated.

Viral Disease

Some viral diseases, such as malignant catarrhal fever and infectious bovine rhinotracheitis, may be confused with pink-eye because they cause cloudiness of the eyeball.

However, affected cattle will also show other signs, related, for example, to the central nervous system or the respiratory system. If you need further advice contact your veterinarian or the District Veterinary Officer at the local DPI office.

TREATMENT

Many cattle recover from pink-eye without treatment, 3-5 weeks after infection. Mustering cattle for the purpose of treating pink-eye may be unwise because dust and flies increase the spread of infection. Consider the percentage of cattle affected before mustering the mob, and potentially spreading the infection to others.

If mustering is needed for essential management procedures such as weaning and drenching, affected cattle should be treated for pink-eye while in the yards, and severely affected cattle, such as those with abscesses in both eyes, should be isolated, treated daily and carefully nursed.

In commercial herds treatment is usually given once only. Treatments options are based on antibiotics to counter the bacteria and anti-inflammatory drugs

ANTIBIOTIC TREATMENTS

There are several proprietary formulations of eye ointments available under prescription from veterinarians, with the active antibiotic constituent Cloxicillin. Application of Cloxicillin based eye ointment into the conjunctival sac (under the upper and lower eyelids) is effective against the Moraxella bacteria. A single application is usually sufficient, but may be repeated every 48 hours in severe infections. Other topical preparations include antibiotic sprays or powders. These only give short term antibiotic coverage and must be repeated several times per day to be effective.

Eye ointments are less irritating and less traumatic to the eye and generally are the recommended treatment. In late and more severe stages of the disease, injection of a combination of a broad-spectrum antibiotic and an anti-inflammatory drug underneath the upper eyelid is often successful. This form of treatment is more expensive, requires head restraint, and is best given by a veterinary surgeon. Occasionally intra-muscular injections of antibiotics and anti-inflammatory drugs are used despite the expense. Applicable with-holding period (WHP) and export slaughter interval (ESI) must be observed with any antibiotic treatments

EYE PATCHES

Eye patches are not a sole treatment for pinkeye, but are a recommended adjunct with an antibiotic preparation. Patches offer protection from further irritation from sunlight, dust and flies. Direct sunlight is painful to a pink-eye affected animal. Patches reduce the pain, so animals are more comfortable and continue to graze, reducing the weight loss often caused by pinkeye.

Denying flies access to the affected eye will help to reduce pinkeye spread within the herd. Patches are glued over the eye, with care taken not to apply glue into the eye.

There are several proprietary eye patches available. Heavy cloth (eg denim) or a dust mask may be a cheap alternative. The patch will fall off after a few weeks.

PREVENTION AND CONTROL

There are some prevention and control measures available..A vaccine has recently been registered for use in Australia. The vaccine, Piliguard®, which works by blocking the attachment of the bacteria to the cornea, will help prevent pinkeye. A single 2ml dose used strategically 3-6 weeks before the onset of the pinkeye season provides significant protection against the pinkeye causing bacteria Moraxella bovis. Annual revaccination is recommended by the manufacturer immediately prior to the beginning of the pinkeye season. Producers may strategically vaccinate the most susceptible lines of stock, namely calves and weaners.

Other control measures include:

- Limiting the spread of the bacteria from animal to animal by controlling fly numbers on cattle with the use of the pour-on synthetic pyrethroid insecticide, deltamethrin. Observe any applicable WHP & ESI;
- Prompt segregation and treatment of any pinkeye in affected stock;
- Avoiding unnecessary yarding of cattle during the problem months.

Experimental work in Queensland has shown that Bos indicus type cattle seem to be less susceptible to pink-eye than British breeds. In British breeds, cattle with complete eyelid pigmentation have been found to be less susceptible to pink-eye.

HAEMOPHILUS DISEASE

Haemophilus is a genus of Gram-negative, pleomorphic, coccobacilli bacteria belonging to the *Pasteurellaceae* family. While *Haemophilus* bacteria are typically small coccobacilli, they are categorised as *pleomorphic* bacteria because of the wide range of shapes they occasionally assume. The genus includes commensal organisms along with some significant pathogenic species such as *H. influenzae*—a cause of sepsis and bacterial meningitis in young children—and

H. ducreyi, the causative agent of chancroid. All members are either aerobic or facultatively anaerobic.

METABOLISM

Members of the *Haemophilus* genus are typically cultured on blood agar plates as all species require at least one of the following blood factors for growth: hemin (factor X) and/or nicotinamide adenine dinucleotide (factor V). Chocolate agar is an excellent *Haemophilus* growth medium as it allows for increased accessibility to these factors. Alternatively, *Haemophilus* is sometimes cultured using the "Staph streak" technique: both *Staphylococcus* and *Haemophilus* organisms are cultured together on a single blood agar plate. In this case, *Haemophilus* colonies will frequently grow in small "satellite" colonies around the larger *Staphylococcus* colonies because the metabolism of *Staphylococcus* produces the necessary blood factor by-products required for *Haemophilus* growth.

LISTERIA

Listeria is a genus of bacteria that contains ten species. Named after the English pioneer of sterile surgery Joseph Lister, the genus received its current name in 1940. *Listeria* species are Gram-positive bacilli. The major human pathogen in the *Listeria* genus is *L. monocytogenes*. It is usually the causative agent of the relatively rare bacterial disease, listeriosis, a serious infection caused by eating food contaminated with the bacteria. The disease affects primarily pregnant women, newborns, adults with weakened immune systems, and the elderly. Listeriosis is a serious disease for humans; the overt form of the disease has a mortality rate of about 20 percent. The two main clinical manifestations are sepsis and meningitis. Meningitis is often complicated by encephalitis, a pathology that is unusual for bacterial infections. *Listeria ivanovii* is a pathogen of mammals, specifically ruminants, and has rarely caused listeriosis in humans.

BACKGROUND

The first documented case of *Listeria* was in 1924. In the late 1920s, two researchers independently identified *Listeria monocytogenes* from animal outbreaks. They proposed the genus *Listerella* in honour of surgeon and early antiseptic advocate Joseph Lister; however, that name was already in use for a slime mold and a protozoan. Eventually, the genus *Listeria* was proposed and accepted. All species within the *Listeria* genus are Gram-positive, non-sporeforming, catalase-positive rods. The genus *Listeria* was classified in the family Corynebacteriaceae through the seventh edition of *Bergey's Manual of Systematic Bacteriology*. The 16S rRNA cataloging studies of Stackebrandt, *et al.* demonstrated that *L. monocytogenes* is a distinct taxon within the

Lactobacillus-Bacillus branch of the bacterial phylogeny constructed by Woese. In 2004, the genus was placed in the newly created Family Listeriaceae. The only other genus in the family is *Brochothrix*.

The genus *Listeria* currently contains ten species: *L. fleischmannii, L. grayi, L. innocua, L. ivanovii, L. marthii, L. monocytogenes, L. rocourtiae, L. seeligeri, L. weihenstephanensis* and *L. welshimeri. Listeria dinitrificans*, previously thought to be part of the *Listeria* genus, was reclassified into the new genus *Jonesia*. Under the microscope, *Listeria* species appear as small, Gram-positive rods, which are sometimes arranged in short chains. In direct smears, they may be coccoid, so they can be mistaken for streptococci. Longer cells may resemble corynebacteria. Flagella are produced at room temperature, but not at 37 °C. Hemolytic activity on blood agar has been used as a marker to distinguish *L. monocytogenes* among other *Listeria* species, but it is not an absolutely definitive criterion. Further biochemical characterisation may be necessary to distinguish between the different species of *Listeria*.

Listeria can be found in soil, which can lead to vegetable contamination. Animals can also be carriers. *Listeria* has been found in uncooked meats, uncooked vegetables, fruit such as cantaloupes, pasteurised or unpasteurised milk, foods made from milk, and processed foods. Pasteurisation and sufficient cooking kill *Listeria*; however, contamination may occur after cooking and before packaging. For example, meat-processing plants producing ready-to-eat foods, such as hot dogs and deli meats, must follow extensive sanitation policies and procedures to prevent *Listeria* contamination.

Listeria monocytogenes is commonly found in soil, stream water, sewage, plants, and food. *Listeria* is responsible for listeriosis, a rare but potentially lethal food-borne infection. The case fatality rate for those with a severe form of infection may approach 25 per cent. (*Salmonella*, in comparison, has a mortality rate estimated at less than 1 per cent.) Although *Listeria monocytogenes* has low infectivity, it is hardy and can grow in temperatures from 4 °C (39.2 °F) (the temperature of a refrigerator), to 37 °C (98.6 °F), (the body's internal temperature).

Listeriosis is a serious illness, and the disease may manifest as meningitis, or affect newborns due to its ability to penetrate the endothelial layer of the placenta.Epidemiology The Center for Science in the Public Interest has published a list of foods that have sometimes caused outbreaks of *Listeria*: hot dogs, deli meats, pasteurised or unpasteurised milk, cheeses (particularly soft-ripened cheeses like feta, Brie, Camembert, blue-veined, or Mexican-style *queso blanco*), raw and cooked poultry, raw meats, ice cream, raw vegetables, and raw and smoked fish. Cantaloupe has been implicated in an outbreak of listeriosis from a farm in Colorado, and the Australian company GMI Food Wholesalers were fined A$236,000 for providing *Listeria monocytogenes*-contaminated chicken wraps to the airline Virgin Blue.

PREVENTION

Preventing listeriosis as a food illness requires effective sanitation of food contact surfaces. Alcohol is an effective topical sanitiser against *Listeria*. Quaternary ammonium can be used in conjunction with alcohol as a food contact safe sanitiser with increased duration of the sanitising action. Refrigerated foods in the home should be kept below 4 °C (39.2 °F) to discourage bacterial growth. Preventing listeriosis also can be done by carrying out an effective sanitation of food contact surfaces.

TREATMENT

In non-invasive listeriosis, the bacteria will often remain within the digestive tract, causing mild symptoms lasting only a few days and requiring only supportive care. Muscle pain and fever in mild cases can be treated with over-the-counter pain relievers, and diarrhea and gastroenteritis can be treated with over-the-counter medications if needed.

In invasive listeriosis, the bacteria has spread to the bloodstream and central nervous system. Treatment includes intravenous delivery of high-dose antimicrobials and in-patient hospital care. Duration of hospital care will vary depending on how widespread the infection is, but is usually no less than 2 weeks. Ampicillin, penicillin, or amoxicillin are often given for invasive listeriosis, and gentamycin is often added in patients with compromised immune systems. Trimethoprim-sulfamethoxazole, vancomycin, and fluoroquinolones can be used in cases of allergy to penicillin. For treatment to be effective, the antibiotic must penetrate the host cell and bind to penicillin-binding protein 3 (PBP3). Cephalosporins are not effective for treatment of listeriosis.

Prompt treatment of listeria infections in pregnancy is critical to prevent the bacteria from infecting the fetus, and antibiotics may be given to pregnant women even in non-invasive listeriosis. These oral therapies in less severe cases can include amoxicillin or erythromycin. In addition to antibiotic therapy, it often recommended that infected pregnant women receive ultrasounds to monitor the health of the fetus. Higher doses of antibiotics are sometimes given to pregnant women to ensure penetration of the umbilical cord and placenta.

Asymptomatic patients who have been exposed to listeria are not recommended for treatment. It is recommended that these patients be informed of the signs and symptoms of the disease and to return for medical care if symptoms present.

RESEARCH

Listeria is an opportunistic pathogen: It is most prevalent in the elderly, pregnant mothers, and AIDS patients. With improved health care leading to a growing elderly population and extended life expectancies for AIDS patients, physicians are more likely to encounter this otherwise-rare infection (only 7 per 1,000,000 healthy people are infected with virulent *Listeria* each year). Better understanding the cell biology of *Listeria* infections, including relevant virulence factors, may lead to better treatments for listeriosis and other intracytoplasmic parasite infections. Researchers are now investigating the use of *Listeria* as a cancer vaccine, taking advantage of its "ability to induce potent innate and adaptive immunity."

LISTERIA MONOCYTOGENES

Listeria monocytogenes is the bacterium that causes the infection listeriosis. It is a facultative anaerobic bacterium, capable of surviving in the presence of oxygen. It can grow and reproduce inside the host's cells and is one of the most virulent food-borne pathogens, with 20 to 30 percent of clinical infections resulting in death. Responsible for approximately 2,500 illnesses and 500 deaths in the United States (U.S.) annually, listeriosis is the leading cause of death among foodborne bacterial pathogens, with fatality rates exceeding even *Salmonella* and *Clostridium botulinum*.

L. monocytogenes is a Gram-positive bacterium, in the division Firmicutes, named for Joseph Lister. Motile via flagella at 30°C and below, but usually not at 37°C, *L. monocytogenes* can instead move within eukaryotic cells by explosive polymerisation of actin filaments (known as comet tails or actin rockets).

Studies suggest up to 10 per cent of human gastrointestinal tracts may be colonised by *L. monocytogenes*.

Nevertheless, clinical diseases due to *L. monocytogenes* are more frequently recognised by veterinarians, especially as meningoencephalitis in ruminants.

Due to its frequent pathogenicity, causing meningitis in newborns (acquired transvaginally), pregnant mothers are often advised not to eat soft cheeses such as Brie, Camembert, feta, and queso blanco fresco, which may be contaminated with and permit growth of *L. monocytogenes*. It is the third-most-common cause of meningitis in newborns.

CLASSIFICATION

L. monocytogenes is a Gram-positive, non-spore-forming, motile, facultatively anaerobic, rod-shaped bacterium. It is catalase-positive and oxidase-negative, and expresses a beta hemolysin, which causes destruction of red blood cells. This bacterium exhibits characteristic tumbling motility when viewed with light microscopy. Although *L. monocytogenes* is actively motile by means of peritrichous flagella at room temperature (20"25°C), the organism does not synthesize flagella at body temperatures (37°C).

The genus *Listeria* belongs to the class, *Bacilli*, and the order, *Bacillales*, which also includes *Bacillus* and *Staphylococcus*. The genus *Listeria* includes seven different species (*L. monocytogenes*, *L. ivanovii*, *L. innocua*, *L. welshimeri*, *L. seeligeri*, *L. grayi*, and *L. murrayi*). Both *L. ivanovii* and *L. monocytogenes* are pathogenic in mice, but only *L. monocytogenes* is consistently associated with human illness. There are 13 serotypes of *L. monocytogenes* that can cause disease, but more than 90 percent of human isolates belong to only three serotypes: 1/2a, 1/2b, and 4b. *L. monocytogenes* serotype 4b strains are responsible for 33 to 50 percent of sporadic human cases worldwide and for all major foodborne outbreaks in Europe and North America since the 1980s.

HISTORY

L. monocytogenes was first described by E.G.D. Murray in 1926 based on six cases of sudden death in young rabbits. Murray referred to the organism as *Bacterium monocytogenes* before Harvey Pirie changed the genus name to *Listeria* in 1940. Although clinical descriptions of *L. monocytogenes* infection in both animals and humans were published in the 1920s, not until 1952 in East Germany was it recognised as a significant cause of neo-natal sepsis and meningitis. Listeriosis in adults would later be associated with patients living with compromised immune systems, such as individuals taking immunosuppressant drugs and corticosteroids for malignancies or organ transplants, and those with HIV infection.

Not until 1981, however, was *L. monocytogenes* identified as a cause of foodborne illness. An outbreak of listeriosis in Halifax, Nova Scotia involving 41 cases and 18 deaths, mostly in pregnant women and neo-nates, was epidemiologically linked to the consumption of coleslaw containing cabbage that had been treated with *L. monocytogenes*-contaminated raw sheep manure. Since then, a number of cases of foodborne listeriosis have been reported, and *L. monocytogenes* is now widely recognised as an important hazard in the food industry.

PATHOGENESIS

Invasive infection by *L. monocytogenes* causes the disease listeriosis. When the infection is not invasive, any illness as a consequence of infection is termed febrile gastroenteritis. The manifestations of listeriosis include septicemia, meningitis (or meningoencephalitis), encephalitis, corneal ulcer, pneumonia, and intrauterine or cervical infections in pregnant women, which may result in spontaneous abortion (second to third trimester) or stillbirth.

Surviving neo-nates of fetomaternal listeriosis may suffer granulomatosis infantiseptica — pyogenic granulomas distributed over the whole body — and may suffer from physical retardation. Influenza-like symptoms, including persistent fever, usually precede the onset of the aforementioned disorders. Gastrointestinal symptoms, such as nausea, vomiting, and diarrhea, may precede more serious forms of listeriosis or may be the only symptoms expressed. Gastrointestinal symptoms were epidemiologically associated with use of antacids or cimetidine.

The onset time to serious forms of listeriosis is unknown, but may range from a few days to three weeks. The onset time to gastrointestinal symptoms is unknown but probably exceeds 12 hours. An early study suggested that *L. monocytogenes* is unique among Gram-positive bacteria in that it might possess lipopolysaccharide, which serves as an endotoxin. Later it was found to not be a true endotoxin. *Listeria* cell walls consistently contain lipoteichoic acids, in which a glycolipid moiety, such as a galactosyl-glucosyl-diglyceride, is covalently linked to the terminal phosphomonoester of the teichoic acid. This lipid region anchors the polymer chain to the cytoplasmic membrane. These lipoteichoic acids resemble the lipopolysaccharides of Gram-negative bacteria in both structure and function, being the only amphipathic polymers at the cell surface.

L. monocytogenes has D-Galactose residues on its surface that can attach to D-Galactose receptors on the host cell walls. These host cells are generally M cells and Peyer's patches of the intestinal mucosa. Once attached to this cells, *L. monocytogenes* can translocate past the intestinal membrane and into the body.

The infective dose of *L. monocytogenes* varies with the strain and with the susceptibility of the victim. From cases contracted through raw or supposedly pasteurised milk, one may safely assume that, in susceptible persons, fewer than 1,000 total organisms may cause disease. *L. monocytogenes* may invade the gastrointestinal epithelium. Once the bacterium enters the host's monocytes, macrophages, or polymorphonuclear leukocytes, it becomes blood-borne (septicemic) and can grow. Its presence intracellularly in phagocytic cells also permits access to the brain and probably transplacental migration to the fetus in pregnant women. The pathogenesis of *L. monocytogenes* centers on its ability to survive and multiply in phagocytic host cells. It seems that Listeria originally evolved to invade membranes of the intestines, as an intracellular infection, and developed a chemical mechanism to do so. This involves a bacterial protein "internalin" which attaches to a protein on the intestinal cell membrane "cadherin". These adhesion molecules are also to be found in two other unusually tough barriers in humans — the blood brain barrier and the feto–placental barrier, and this may explain the apparent affinity that Listeria has for causing meningitis and affecting babies in-utero.

REGULATION OF PATHOGENESIS

L. monocytogenes can act as a saprophyte or a pathogen, depending on its environment. When this bacterium is present within a host organism, quorum sensing causes the up regulation of several virulence genes. Depending on the location of the bacterium within the host organism, different activators up regulate the virulence genes. SigB, an alternative sigma factor, up regulates Vir genes in the intestines, whereas PrfA up regulates gene expression when the bacterium is present in blood. Little is known about how this bacterium switches between acting as a saprophyte and a pathogen; however, several non-coding RNAs are thought to be required to induce this change.

PATHOGENICITY OF LINEAGES

L. monocytogenes has three distinct lineages, with differing evolutionary histories and pathogenic potentials. Lineage I strains contain the majority of human clinical isolates and all human epidemic clones, but are underrepresented in animal clinical isolates. Lineage II strains are overrepresented in animal cases and underrepresented in human clinical cases, and are more prevalent in environmental and food samples. Lineage III isolates are very rare, but significantly more common in animal than human isolates.

DETECTION

Anton Test:A test used in the identification of Listeria monocytogenes; instillation of a culture into the conjunctival sac of a rabbit or guinea pig causes severe keratoconjunctivitis within 24 hours. Culture Charecteristics:Listeria grows on media such as Mueller-Hinton agar. Identification is enhanced if the primary cultures are done on agar containing sheep blood, because the characteristic small zone of hemolysis can be observed around and under colonies. Isolation can be enhanced if the tissue is kept at 4 °C for some days before inoculation into bacteriologic media. The organism is a facultative anaerobe and is catalase-positive and motile. Listeria produces acid but not gas in a variety of carbohydrates. The motility at room temperature and hemolysin production are primary findings that help differentiate listeria from coryneform bacteria.

The methods for analysis of food are complex and time-consuming. The present U.S. FDA method, revised in September 1990, requires 24 and 48 hours of enrichment, followed by a variety of other tests. Total time to identification takes from five to seven days, but the announcement of specific non-radiolabelled DNA probes should soon allow a simpler and faster confirmation of suspect isolates.

Recombinant DNA technology may even permit two- to three-day positive analysis in the future. Currently, the FDA is collaborating in adapting its methodology to quantitate very low numbers of the organisms in foods.

TREATMENT

When listeric meningitis occurs, the overall mortality may reach 70 per cent, from septicemia 50 per cent, and from perinatal/neo-natal infections greater than 80 per cent. In infections during pregnancy, the mother usually survives. Reports of successful treatment with parenteral penicillin or ampicillin exist. Trimethoprim-sulfamethoxazole has been shown effective in patients allergic to penicillin.

A bacteriophage, *Listeria phage P100*, has been proposed as food additive to control *Listeria monocytogenes*. Bacteriophage treatments have been developed by several companies. EBI Food Safety and Intralytix both have products suitable for treatment of the bacterium. The U.S. Food and Drug Administration (FDA) approved a cocktail of six bacteriophages from Intralytix, and a one type phage product from EBI Food Safety designed to kill *L. monocytogenes*. Uses would potentially include spraying it on fruits and ready-to-eat meat such as sliced ham and turkey.

USE AS A TRANSFECTION VECTOR

Because *L. monocytogenes* is an intracellular bacterium, some studies have used this bacterium as a vector to deliver genes *in vitro*. Current transfection efficiency remains poor. One example of the successful use of *L. monocytogenes* in *in vitro* transfer technologies is in the delivery of gene therapies for cystic fibrosis cases.

CANCER VACCINE

A live attenuated *L. monocytogenes* cancer vaccine, ADXS11-001, is under development as a possible treatment for cervical carcinoma.

EPIDEMIOLOGY

Researchers have found *L. monocytogenes* in at least 37 mammalian species, both domesticated and feral, as well as in at least 17 species of birds and possibly in some species of fish and shellfish. Laboratories can isolate *L. monocytogenes* from soil, silage, and other environmental sources. *L. monocytogenes* is quite hardy and resists the deleterious effects of freezing, drying, and heat remarkably well for a bacterium that does not form spores. Most *L. monocytogenes* strains are pathogenic to some degree.

ROUTES OF INFECTION

L. monocytogenes has been associated with such foods as raw milk, pasteurised fluid milk, cheeses (particularly soft-ripened varieties), ice cream, raw vegetables, fermented raw-meat sausages, raw and cooked poultry, raw meats (of all types), and raw and smoked fish. Its ability to grow at temperatures as low as 0°C permits multiplication in refrigerated foods. At refrigeration temperature, such as 4°C, the amount of ferric iron can affect the growth of *L. monocytogenes*.

INFECTIOUS CYCLE

The primary site of infection is the intestinal epithelium, where the bacteria invade non-phagocytic cells via the "zipper" mechanism. Uptake is stimulated by the binding of listerial internalins (Inl) to E-cadherin, a host cell adhesion factor, or Met (c-Met), hepatocyte growth factor. This binding activates certain Rho-GTPases, which subsequently bind and stabilise Wiskott Aldrich syndrome protein (WAsp). WAsp can then bind the Arp2/3 complex and serve as an actin nucleation point. Subsequent actin polymerisation creates a "phagocytic cup", an actin-based structure normally formed around foreign materials by phagocytes prior to endocytosis. The net effect of internalin binding is to exploit the junction-forming apparatus of the host into internalising the bacterium. *L. monocytogenes* can also invade phagocytic cells (*e.g.*, macrophages), but requires only internalins for invasion of non-phagocytic cells.

Following internalisation, the bacterium must escape from the vacuole/ phagosome before fusion with a lysosome can occur. Three main virulence factors that allow the bacterium to escape are listeriolysin O (LLO-encoded by *hly*) phospholipase A (encoded by *plcA*) and phospholipase B (*plcB*). Secretion of LLO and PlcA disrupts the vacuolar membrane and allows the bacterium to escape into the cytoplasm, where it may proliferate.

Once in the cytoplasm, *L. monocytogenes* exploits host actin for the second time. ActA proteins associated with the old bacterial cell pole (being a bacillus, *L. monocytogenes* septates in the middle of the cell and thus has one new pole and one old pole) are capable of binding the Arp2/3 complex, thereby inducing actin nucleation at a specific area of the bacterial cell surface. Actin polymerisation then propels the bacterium unidirectionally into the host cell membrane. The protrusion that is formed may then be internalised by a neighbouring cell, forming a double-membrane vacuole from which the bacterium must escape using LLO and PlcB. This mode of direct cell-to-cell spread involves a cellular mechanism known as paracytophagy.

ERYSIPELOTHRIX

CLASSIFICATION

Higher Order Taxa

- Bacteria;
- Firmicutes;
- Mollicutes;
- Anaeroplasmatales;
- Erysipelotrichidae;
- *Erysipelothrix*

Species

- Erysipelothrix rhusiopathiae,
- Erysipelothrix tonsillarum,
- *Erysipelothrix inopinata*

Description and Significance

Erysipelothrix are pathogenic bacteria that infect over 50 animal species, but are most commonly found in domesticated pigs. There are three species: *Erysipelothrix rhusiopathiae, E. Tonsillarum*, and *E. inopinata*, each of which have a smooth strain and a rough strain which vary in virulence. The smooth strain of each species is pathogenic, while the rough is not. *Erysipelothrix rhusiopathiae* was first isolated by Koch in 1876. This was the only known species until a subsequent species was discovered in the tonsils of apparently healthy pigs, which was named *E. Tonsillarum*. Then a third species, *Erysipelothrix inopinata* was first isolated in the course of sterile filtration of vegetable peptone broth.

Genome Structure

The USDA Microbial Genomics Stakeholder Workshop for Animal Health and Food Safety Pathogens considered sequencing *Erysipelothrix rhusiopathiae* as a swine pathogen. However, the organisation chose another organism to sequence, and there are no current plans to sequence *Erysipelothrix rhusiopathiae.*

Cell Structure and Metabolism

Erysipelothrix are chemoorganotrophic facultative anaerobes with a respiratory metabolism and are weakly fermentative. *Erysipelothrix* are gram positive but may appear gram negative because they decolourise easily. Their cells are catalase-negative and oxidase negative.

Erysipelothrix are non-encapsulated, non-sporulating and non-motile. Their slender rods have rounded ends and are straight or slightly curved, but tend to form long filaments. The cell wall is what helps to differentiate these bacteria from others with a B-cell wall type because the peptide bridge is formed between amino acids at positions 2 and 4 of adjacent peptide side chains. This distinguishes *Erysipelothrix* from other bacteria because most others have the peptide bridge formed between amino acids at positions 3 and 4.

Colonies of these bacteria have two distinct forms which are the smooth and a larger rough form. The smooth is about.1mm in diameter, convex and circular. The rough form is about.2-.4 mm in diameter, flat with a matte surface. Organisms are either arranged in single short chains, in pairs as a "V" configuration or are grouped randomly.

The exact growth requirements of the organism have not yet been determined. However, several amino acids, riboflavin, and small amounts of oleic acid are required, and growth is enhanced by tryptophan. Growth occurs at an optimal temperature of 30-37°C, a pH of 7.2-7.6 and is improved by 5-10 per cent carbon dioxide.

Ecology

Erysipelothrix rhusiopathiae can be found in sewage, the guts of fish and ground contaminated with animal feces. These bacteria are unable to survive for a long period of time in external environments since the rough form prefers slightly acidic conditions at 37°C and the smooth prefers a slightly alkaline environment at 30°C.

Epidemiology

Since the late 1800's *Erysipelothrix rhusiopathiae* has been recognised as a cause of infections in both domesticated animals and humans. It has been the source of many diseases in turkey, chicken, ducks, emus, and sheep. Most importantly, it is the etiological agent of swine *erysipelas*. Most human cases are the result of the occupational hazards of working with animals and animal waste products. Recently, it has been suggested that incidence of human infection could be declining due to technological advances in animal industries. But infection still occurs in certain environments such as Japan, where there are animal hygiene problems, and Western Australia where there are many lobster fisherman and handlers.

Infection by this organism is often misdiagnosed because of difficulty in isolation and identification and its close resemblance to other infections. The bacteria resides deep in the skin, making culturing it a lengthy process. Due to the difficulty in culturing, diagnosis is additionally complicated. However, blood samples or tissue biopsies can be taken in order to isolate the organism more easily for diagnosis of disease.

There have been many recent studies leading to advances in molecular approaches to the diagnosis of an *E. rhusiopathiae* infection. Understanding of *Erysipelothrix rhusiopathiae* taxonomy and pathogenesis has also been greatly improved by recent studies. Although antibiotic treatments such as ampicillin and penicillin have been discovered to be effective, it has been shown that containment and control procedures are a far more effective way to reduce infection in both humans and animals.

Pathology

Smooth strains of *Erysipelothrix* are pathogenic and usually enter their host through scratches or puncture wounds on the surface of the skin, while rough strains are non-pathogenic. Swine *Erysipelothrix*, called *Erysipelas*, has four forms; an acute form, a subacute uriticarial form, a chronic non-supportive arthritic form, and a chronic cardiac form. The acute form causes septicemia, fever, anorexia, diarrhea, cyanosis and death. The subacute uriticarial form causes diamond-shaped skin lesions, alopecia, sloughing of tail tip and ear tips, and hyperkeratosis. There are three forms of *Erysipelothrix* in humans, called *Erysipeloid,* and they are usually occupationally related. The first is a localised cutaneous form whose symptoms include a throbbing, itching pain and swelling of the finger or hand. The second is a generalised cutaneous form and the third a septicemic form which is associated with the heart disease endocarditis.

Erysipelothrix causes septicemia and eventually death in some wild birds. Although it lives on the skin of fish, they are not infected with the disease. These bacteria also cause joint illness in sheep, lamb, and cattle.

SWINE BRUCELLOSIS

Swine brucellosis is a zoonosis affecting pigs, caused by the bacteria *Brucella suis*. The disease typically causes chronic inflammatory lesions in the reproductive organs of susceptible animals or orchitis and may even affect joints and other organs. The most common symptom is abortion in pregnant susceptible hosts at any stage of gestation. Other manifestations are temporary or permanent sterility, lameness, posterior paralysis, spondylitis, and abscess formation. It is transmitted mainly by ingestion of infected tissues or fluids, semen during breeding, and suckling infected animals. In humans, it can cause undulant fever.

Since brucellosis threatens food supply and cause undulant fever, *Brucella suis* and other *Brucella* species (*B. melitensis, B. abortis, B. ovis, B. canis*) are recognised as potential agricultural, civilian, and military bioterrorism agents.

ETIOLOGY

B. suis are gram-negative, facultative intracellular coccobacilli and therefore are capable of growing and reproducing inside of host cells, specifically phagocytic cells. They are also non-spore-forming, non-capsulated, and non-motile. Flagellar genes, however, are present in the *B. suis* genome, but thought to be cryptic remnants because some were truncated and others were missing crucial components of the flagellar apparatus. Interestingly, in mouse models, studies have shown that the flagellum is essential for a normal infectious cycle, where the inability to assemble a complete flagellum leads to severe attenuation of the bacteria.

Brucella suis are differentiated into five biovars (strains), where bv. 1-3 infect boars and pigs and bv.1 and 3 may cause severe diseases in humans. In contrast, bv. 2 found in wild boars in Europe show mild or no clinical signs and cannot infect healthy humans, but do infect pigs and hares.

PATHOGENESIS

Phagocytes are an essential component of the host's innate immune system with various anti-microbial defence mechanisms to clear pathogens via oxidative burst, acidificiation of phagosomes, and fusion of the phagosome and lysosome. *B. suis*, in retun, have developed ways to counteract the host cell defence to survive in the macrophage and to deter host immune responses.

Brucella suis possess smooth lipopolysaccharide (LPS), which have a full length O-chain, as opposed to rough LPS, which have a truncated or no O-chain. This structural characteristic allows for *B. suis* to interact with lipid rafts on the surface of macrophages to be internalised, and the formed lipid rich phagosome is able to avoid fusion with lysosomes through this endocytic pathway. In addition, this furtive entry into the macrophage does not affect the cell's normal trafficking. The smooth LPS also inhibits host cell apoptosis via O-polysaccharides through a TNF-alpha independent mechanism, which allows for *B. suis* to avoid the activation of the host immune system.

Once inside the macrophage, *B. suis* are able to endure the rapid acidificiation in the phagosome to pH 4.0-4.5 by expressing metabolism genes mainly for amino acid synthesis. The acidic pH is actually essential for replication of the bacteria by inducing major virulence genes of the virB operon and the synthesis of DnaK chaperones. DnaK is part of the heat shock protein 70 family and aids in the correct synthesis and activation of certain virulence factors.

In addition, the *B. suis* gene for nickel transport, *nikA*, is activated by metal ion deficiency and is expressed once in the phagosome. Nickel is essential for many enzymatic reactions including ureolysis to produce ammonia which in turn may neutralise acidic pH. It is suggested that since *B. suis* is unable to grow in strongly acidic medium, it could be protected from acidification by the ammonia.

SUMMARY

- *B. suis* encounters macrophage, but no oxidative burst occurs
- Lipid rafts are necessary for macrophage penetration
- Phagosome rapidly acidifies creating stressful environment for bacteria which triggers activation of virulence genes
- Lipid rafts on phagosomes prevent lysosomal fusion and normal cell trafficking is unaffected

SYMPTOMS

The most frequent clinical sign following *Brucella suis* infection is abortion in pregnant females, reduced milk production, and infertility. Cattle can also be transiently infected when they share pasture or facilities with infected pigs and *B. suis* can be transmitted by cow's milk. Swine also develop orchitis (swelling of the testicles), lameness (movement disability), hind limb paralysis, or spondylitis (inflammation in joints).

TREATMENT

Because *Brucella suis* are facultative intracellular and are able to adapt to environmental conditions in the macrophage, treatment failure and relapse rates are high. The only effective way to control and eradicate zoonosis is by vaccination of all susceptible hosts and elmination of infected animals. The *Brucella abortus* (rough LPS *Brucella*) vaccine, developed for bovine brucellosis is licensed by the USDA Animal Plant Health Inspection Service, has shown protection for some swine and is also effective against *B. suis* infection, however, there is currently no approved vaccine for swine brucellosis.

BIOLOGICAL WARFARE

In the United States, *Brucella suis* was the first biological agent weaponised in 1952 and was field-tested with *B. suis*-filled bombs called M33 cluster bomb. It is, however, considered to be one of the agents of lesser threat because many infections are asymptomatic and the mortality is low, but it is used more as an incapacitating agent.

PASTEURELLACEAE OF BACTERIA

Pasteurellaceae comprise a large and diverse family of Gram-negative Proteobacteria with members ranging from important pathogens such as *Haemophilus influenzae* to commensals of the animal and human mucosa. Most members live as commensals on mucosal surfaces of birds and mammals, especially in the upper respiratory tract. The family includes several pathogens of vertebrates, most notably *H. influenzae*. This species causes several diseases in humans (though not flu, as was originally thought). Other *Pasteurellaceae* cause gingivitis and chancroid in humans and many others are important veterinary pathogens.

Pasteurellaceae are typically rod-shaped, and are a notable group of facultative anaerobes. They can be distinguished from the related Enterobacteriaceae by the presence of oxidase, and from most other similar bacteria by the absence of flagella.

Bacteria in the family Pasteurellaceae have been classified into a number of genera based on metabolic properties, but these classifications are not generally accurate reflections of the evolutionary relationships between different species. *H. influenzae* was the first organism to have its genome sequenced and has been studied intensively by genetic and molecular methodologies. Since 1995, the family has been expanded from three genera to the current thirteen through the use of new genetic-based classification and identification technologies. Many members of the *Pasteurellaceae* family make excellent natural models for the study of bacterial pathogenesis and host-pathogen-interactions thus giving valuable insights into related human diseases.

TAXONOMY AND BIODIVERSITY

The family *Pasteurellaceae* includes 38 properly classified species in addition to 24 misclassified species. The majority of taxa have been isolated from disease conditions in warm blooded animals and in particular in farm animals. These bacteria are obligate parasites or commensals of vertebrates, colonising mainly the mucosal surfaces of the upper respiratory tract, oropharynx, and reproductive tracts and possibly also parts of the intestinal tract. Most taxa represent potential pathogens. Both systemic and local infections have been reported for most taxa involved in diseases. However, pneumonia has been reported most frequently. Fossil remnants of members of *Pasteurellaceae* have never been reported and information on the diversification of taxa within the family can only be obtained by phylogenetic reconstruction.

MOLECULAR SIGNATURES

Comparative analyses of *Pasteurellaceae* genomes have identified large numbers (>20) of conserved signature indels (CSIs) in different important proteins that are uniquely shared by all sequenced *Pasteurellaceae* species/ strains, but are not found in any other bacteria. Based upon many other conserved indels that are specific for subgroups of *Pasteurellaceae* species, it has been proposed to divide the family *Pasteurellaceae* into at least two clades. One proposed clade includes *Aggregatibacter*, *Pasteurella*, *Actinobacillus succinogenes*, *Haemophilus influenzae*, *Haemophilus somnus* and *Mannheimia succiniciproducens*; while the other includes *Actinobacillus minor*, *Actinobacillus pleuropneumoniae*, *Haemophilus ducryi*, *Haemophilus parasuis* and *Mannheimia haemolytica*.

RTX TOXIN

RTX toxins are bacterial pore forming toxins that are particularly abundant among pathogenic species of *Pasteurellaceae* where they play a major role in virulence. RTX toxins of several primary pathogens of the family of *Pasteurellaceae* are directly involved in causing necrotic lesions of the target organs. Many RTX toxins are mainly known as haemolysins due to their capacity to lyse erythrocytes (red blood cells) *in vitro*, an effect that seems to be non-specific. It is now known for many RTX toxins that their specific targets are leukocytes, where RTX toxins bind to the corresponding b subunit (CD18) of b_2 integrons and then cause a cytotoxic effect. For several RTX toxins the binding to CD18 was shown to be host specific and seems to be the basis determining the host range of a given RTX toxin. Observations on very closely related species of the *Pasteurellaceae* family with different RTX toxins indicate that these latter contribute to a significant part to the host specificity of the pathogen itself. RTX toxins induce a strong immunologic response generating neutralising antibodies. They therefore constitute important antigens in modern subunit vaccines.

IRON UPTAKE

Outer membrane (OM) proteins for iron acquisition have roles in infection and pathogenesis and have growing appeal as novel targets for anti-infectives and therapeutics. Characterisation of cell surface proteins of members of the *Pasteurellaceae* family including *Haemophilus*, *Actinobacillus*, *Pasteurella*, and the *Mannheimia* genera of organisms has highlighted several redundant iron acquisition receptors for transferrin, siderophores, and heme/heme-containing proteins. In addition, the identification of several immunogenic lipoproteins and OM proteins has driven research for an effective cross-protective vaccine for these organisms.

DNA SEQUENCE

DNA sequence data are available for the following *Pasteurellaceae* members: *Aggregatibacter actinomycetemcomitans* strain HK1651, *Actinobacillus pleuropneumoniae* strains L20 and sv1 4074, [*Haemophilus*] *ducreyi* strain 35000HP, *Haemophilus influenzae* strains 86-028NP, R2846, R2866, and Rd, *Histophilus somni* strains 129Pt and 2336, *Mannheimia haemolytica* A1 strain ATCC BAA-410, *Mannheimia succiniciproducens* strain MBEL55E, and *Pasteurella multocida* strain Pm70.

HAEMOPHILUS INFLUENZAE

Haemophilus influenzae, formerly called Pfeiffer's bacillus or *Bacillus influenzae*, is a Gram-negative, coccobacilli bacterium first described in 1892 by Richard Pfeiffer during an influenza pandemic. A member of the *Pasteurellaceae* family, it is generally aerobic, but can grow as a facultative anaerobe. *H. influenzae* was mistakenly considered to be the cause of influenza until 1933, when the viral etiology of influenza became apparent. The bacterium is colloquially known as *bacterial influenza*. Still, *H. influenzae* is responsible for a wide range of clinical diseases. *H. influenzae* was the first free-living organism to have its entire genome sequenced. The sequencing project was completed and published in 1995.

SEROTYPES

In 1930, two major categories of *H. influenzae* were defined: the unencapsulated strains and the encapsulated strains. Encapsulated strains were classified on the basis of their distinct capsular antigens. There are six generally recognised types of encapsulated *H. influenzae*: a, b, c, d, e, and f.

Genetic diversity among unencapsulated strains is greater than within the encapsulated group. Unencapsulated strains are termed non-typable (NTHi) because they lack capsular serotypes; however, they can be classified by multilocus sequence typing. The pathogenesis of *H. influenzae* infections is not completely understood, although the presence of the capsule in encapsulated type b (Hib), a serotype causing conditions such as epiglottitis, is known to be a major factor in virulence. Their capsule allows them to resist phagocytosis and complement-mediated lysis in the non-immune host. The unencapsulated strains are almost always less invasive; they can, however, produce an inflammatory response in humans, which can lead to many symptoms. Vaccination with Hib conjugate vaccine is effective in preventing Hib infection, but does not prevent infection with NTHi strains.

DISEASES

Table. Haemophilus Influenzae Infection.

Classification and External Resources	
ICD-10	A49.2
ICD-9	041.5
DiseasesDB	5570
MedlinePlus	000612 (Meningitis)
eMedicine	topic list

Most strains of *H. influenzae* are opportunistic pathogens; that is, they usually live in their host without causing disease, but cause problems only when other factors (such as a viral infection, reduced immune function or chronically inflamed tissues, *e.g.*, from allergies) create an opportunity. They infect the host by sticking to the host cell using Trimeric Autotransporter Adhesins (TAA).

Naturally acquired disease caused by *H. influenzae* seems to occur in humans only. In infants and young children, *H. influenzae* type b (Hib) causes bacteremia, pneumonia, and acute bacterial meningitis. On occasion, it causes cellulitis, osteomyelitis, epiglottitis, and infectious arthritis.

In fact, *Haemophilus influenzae* is the most common etiologic agent associated with epiglottitis (thumbprint sign seen on X-Ray).

Due to routine use of the Hib conjugate vaccine in the U.S., since 1990, the incidence of invasive Hib disease has decreased to 1.3/100,000 in children. However, Hib remains a major cause of lower respiratory tract infections in infants and children in developing countries where the vaccine is not widely used.

Unencapsulated *H. influenzae* strains are unaffected by the Hib vaccine and cause ear infections (otitis media), eye infections (conjunctivitis), and sinusitis in children, and are associated with pneumonia.

DIAGNOSIS

Haemophilus influenzae requires X and V factors for growth. In this culture haemophilus has only grown around the paper disc that has been impregnated with X and V factors. There is no bacterial growth around the discs that only contain either X or V factor. Clinical diagnosis of *H. influenzae* is typically performed by bacterial culture or latex particle agglutinations. Diagnosis is considered confirmed when the organism is isolated from a sterile body site. In this respect, *H. influenzae* cultured from the nasopharyngeal cavity or sputum would not indicate *H. influenzae* disease, because these sites are colonised in disease-free individuals. However, *H. influenzae* isolated from cerebrospinal fluid or blood would indicate *H. influenzae* infection.

CULTURE

Bacterial culture of *H. influenzae* is performed on agar plates, the preferable one being chocolate agar, with added X (hemin) and V (nicotinamide adenine dinucleotide) factors at 37°C in a CO_2-enriched incubator. Blood agar growth is only achieved as a satellite phenomenon around other bacteria. Colonies of *H. influenzae* appear as convex, smooth, pale, grey or transparent colonies.

Gram-stained and microscopic observation of a specimen of *H. influenzae* will show Gram-negative, rod shaped, with no specific arrangement. The cultured organism can be further characterised using catalase and oxidase tests, both of which should be positive. Further serological testing is necessary to distinguish the capsular polysaccharide and differentiate between *H. influenzae* b and non-encapsulated species.

Although highly specific, bacterial culture of *H. influenzae* lacks in sensitivity. Use of antibiotics prior to sample collection greatly reduces the isolation rate by killing the bacteria before identification is possible. Beyond this, *H. influenzae* is a finicky bacterium to culture, and any modification of culture procedures can greatly reduce isolation rates. Poor quality of laboratories in developing countries has resulted in poor isolation rates of *H. influenzae*.

H. influenzae will grow in the hemolytic zone of *Staphylococcus aureus* on blood agar plates; the hemolysis of cells by *S. aureus* releases factor V which is needed for its growth. *H. influenzae* will not grow outside the hemolytic zone of *S. aureus* due to the lack of nutrients such as factor V in these areas. Fildes agar is best for isolation. In Levinthal medium capsulated strains show distinctive iridescence.

LATEX PARTICLE AGGLUTINATION

The latex particle agglutination test (LAT) is a more sensitive method to detect *H. influenzae* than culture. Because the method relies on antigen rather than viable bacteria, the results are not disrupted by prior antibiotic use. It also has the added benefit of being much quicker than culture methods. However, antibiotic sensitivity testing is not possible with LAT alone, so a parallel culture is necessary.

MOLECULAR METHODS

Polymerase chain reaction (PCR) assays have been proven to be more sensitive than either LAT or culture tests, and highly specific. However, PCR assays have not yet become routine in clinical settings. Countercurrent immunoelectrophoresis has been shown to be an effective research diagnostic method, but has been largely supplanted by PCR.

INTERACTION WITH STREPTOCOCCUS PNEUMONIAE

Both *H. influenzae* and *S. pneumoniae* can be found in the upper respiratory system of humans. In an *in vitro* study of competition, *S. pneumoniae* always overpowered *H. influenzae* by attacking it with hydrogen peroxide and stripping off the surface molecules *H. influenzae* needs for survival.

When both bacteria are placed together into a nasal cavity, within 2 weeks, only *H. influenzae* survives. When either is placed separately into a nasal cavity, each one survives. Upon examining the upper respiratory tissue from mice exposed to both bacteria species, an extraordinarily large number of neutrophils (immune cells) was found. In mice exposed to only one bacterium, the cells were not present.

Lab tests showed neutrophils exposed to dead *H. influenzae* were more aggressive in attacking *S. pneumoniae* than unexposed neutrophils. Exposure to dead *H. influenzae* had no effect on live *H. influenzae*.

Two scenarios may be responsible for this response:

1. When *H. influenzae* is attacked by *S. pneumoniae*, it signals the immune system to attack the *S. pneumoniae*
2. The combination of the two species triggers an immune system response that is not set off by either species individually.

It is unclear why *H. influenzae* is not affected by the immune response.

TREATMENT

Haemophilus influenzae produces beta-lactamases, and it is also able to modify its penicillin-binding proteins, so it has gained resistance to the penicillin family of antibiotics. In severe cases, cefotaxime and ceftriaxone delivered directly into the bloodstream are the elected antibiotics, and, for the less severe cases, an association of ampicillin and sulbactam, cephalosporins of the second and third generation, or fluoroquinolones are preferred. (Fluoroquinolone-resistant Haemophilus influenzae has been observed.)

Macrolide antibiotics (*e.g.*, clarithromycin) may be used in patients with a history of allergy to beta-lactam antibiotics. Macrolide resistance has also been observed.

PREVENTION

Effective vaccines for *Haemophilus influenzae* Type B have been available since the early 1990s, and is recommended for children under age 5 and asplenic patients. The World Health Organisation recommends a pentavalent vaccine, combining vaccines against diphtheria, tetanus, pertussis, hepatitis B and Hib. There is not yet sufficient evidence on how effective this pentavalent vaccine is in relation to the individual vaccines. Hib vaccines cost about seven times the total cost of vaccines against measles, polio, tuberculosis, diphtheria, tetanus, and pertussis. Consequently, whereas 92 per cent of the populations of developed countries was vaccinated against Hib as of 2003, vaccination coverage was 42 per cent for developing countries, and only 8 per cent for least-developed countries.

SEQUENCING

H. influenzae was the first free-living organism to have its entire genome sequenced. Completed by Craig Venter and his team, *Haemophilus* was chosen because one of the project leaders, Nobel laureate Hamilton Smith, had been working on it for decades and was able to provide high-quality DNA libraries. The genome consists of 1,830,140 base pairs of DNA in a single circular chromosome that contains 1740 protein-coding genes, 2 transfer RNA genes, and 18 other RNA genes. The sequencing method used was whole-genome shotgun, which was completed and published in *Science* in 1995 and conducted at The Institute for Genomic Research.

LIKELY PROTECTIVE ROLE OF TRANSFORMATION

Unencapsulated *H. influenzae* is often observed in the airways of patients with chronic obstructive pulmonary disease (COPD). Neutrophils are also observed in large numbers in sputum from patients with COPD. The neutrophils phagocytise *H. influenzae*, thereby activating an oxidative respiratory burst. However instead of killing the bacteria the neutrophils are themselves killed (though such an oxidative burst likely causes DNA damage in the *H. influenzae* cells). The lack of killing of the *H. influenzae* appears to explain the persistence of infection in COPD. *H. influenzae* mutants defective in the *rec1* gene (a homolog of *recA*) are very sensitive to killing by the oxidising agent hydrogen peroxide. This finding suggests that *rec1* expression is important for *H. influenzae* survival under conditions of oxidative stress. Since it is a homolog of *recA*, *rec1* likely plays a key role in recombinational repair of DNA damage. Thus *H. influenzae* may protect its genome against the reactive oxygen species produced by the host's phagocytic cells through recombinational repair of oxidative DNA damages. Recombinational repair of a damaged site of a chromosome requires, in addition to *rec1*, a second homologous undamaged DNA molecule. Individual *H. influenzae* cells are capable of taking up homologous DNA from other cells by the process of transformation. Transformation in *H. influenzae* involves at least 15 gene products, and is likely an adaptation for repairing DNA damages in the resident chromosome (as suggested in Transformation (genetics)#Transformation, as an adaptation for DNA repair).

PASTEURELLA AND MANNHEIMIA PNEUMONIAS

Bronchopneumonia caused by Pasteurella multocida or Mannheimia haemolytica has a cranioventral lung distribution and affects sheep and goats of all ages worldwide. It can be particularly devastating in young animals. It is a common cause of morbidity and mortality in lambs and kids, especially in those that have not received adequate colostrum or in which passive colostral immunity is waning. The disease appears to occur most often in animals that have undergone recent stress such as transportation, weaning, or commingling with animals from unrelated farms.

ETIOLOGY

P multocida and M haemolytica are gram-negative rods that can cause pneumonia either alone or in conjunction with other organisms. Primary infections with respiratory pathogens such as parainfluenza-type 3, adenovirus, respiratory syncytial virus, Bordetella parapertussis, or Mycoplasma spp appear to predispose to secondary infection with Pasteurella and Mannheimia. Both organisms are normal inhabitants of the upper respiratory tract of sheep and goats; however, pneumonia caused by M haemolytica biotype A occurs more frequently than infection with P multocida. Similar to the infection in calves, P multocida in sheep and goats produces a milder disease with a shorter course than M haemolytica-associated pneumonia. In sheep and goats, serotype A_2 is the most common serotype isolated. Important but less prevalent serotypes found in sheep are A_1, A_6, A_7, A_8, A_9, A_{12}, and in goats, serotypes A_1 and A_6.

PATHOGENESIS

Stress appears to be an important factor in the breakdown of respiratory defence mechanisms, allowing Pasteurella, Mannheimia, Mycoplasma spp, other bacteria, and viruses to invade lung tissue and cause pneumonia. In some laboratory animal species and calves, alveolar macrophage function is impaired after viral pneumonia. This results in decreased clearance of inhaled bacterial pathogens, allowing them to become established. Pathogen-host interactions result in tissue damage, especially because of massive influx of neutrophils. As these neutrophils are lysed, enzymes are released that cause more lung tissue damage. This mechanism may be similar to that of Pasteurella and Mannheimia pneumonias in sheep and goats.

CLINICAL FINDINGS

M haemolytica-associated pneumonia occurs in both young stock and mature animals, usually during times of stress such as lambing/kidding, weaning, dramatic changes in environmental temperatures, inadequate ventilation situations, and less than ideal stocking densities, especially in confinement housing. Outbreaks in groups of sheep and goats usually occur 10–14 days after a stress. For example, in feedlots, outbreaks are expected ~2 wk after arrival in the feedlot.

Early clinical signs can be dramatic and reflect the endotoxemia that often occurs with infection. Sudden death may occur without observation of clinical signs. The disease usually has a rapid onset accompanied by respiratory distress, including open-mouth breathing, anorexia, depression, poor perfusion (prolonged capillary refill time and cool extremities), fever of 104–106°F (40–41.1°C), serous (early) to mucopurulent (later) ocular and nasal discharges, coughing, and lethargy. Harsh lung sounds, especially in the cranioventral portions of the lung field, may be auscultated. Morbidity and mortality rates are variable. Animals that survive the acute phase may develop secondary problems that result in chronic ill health.

LESIONS

Lesions are usually confined to the cranioventral lung lobes on both sides. These areas may appear red to purple and feel firm from consolidation. The pleural cavity may contain variable amounts of straw-coloured fluid, and yellow fibrin may cover the pleural surface of affected lung lobes from pleuritis. Chronic cases may have extensive pleural adhesions and multiple abscesses of variable size.

DIAGNOSIS

In acute cases, cultures obtained from tracheal swabs or washes or from lung tissue or associated lymph nodes are diagnostic. Histopathologic examination is useful, especially if other types of pneumonia (eg, retrovirus interstitial pneumonia in adult sheep and goats) are also suspected. In chronic cases, bacterial cultures may be less rewarding; Pasteurella or Mannheimia may have been the initial problem, but results of cultures taken later may reveal Arcanobacterium pyogenes, a common causative agent of lung abscesses.

TREATMENT AND CONTROL

Whenever possible, treatment should be based on bacterial culture and sensitivity, especially in herd or flock outbreaks, when valuable animals are involved, or in acute or chronic cases when initial therapeutic attempts have failed. Commonly recommended antibiotics include ceftiofur (1.1–2.2 mg/kg, sid), oxytetracycline (10 mg/kg, sid, of non-long-acting product, or 20 mg/kg once of the long-acting product), ampicillin (20 mg/kg, bid), flor-fenicol (20 mg/kg, every 48 hr), and tylosin (10–20 mg/kg, sid-bid). Therapy should continue for at least 24–48 hr after body temperature has returned to normal. Duration of treatment usually is 4–5 days. Acute cases may also benefit from the use of NSAID (eg, aspirin, flunixin meglumine, or ketoprofen) in conjunction with antibiotic therapy for control of endotoxemia and inflammation. Treatment with NSAID should be of short duration because prolonged use may result in gastric ulceration or renal complications. In the USA, use of some of the above antibiotics and NSAID is extra-label, and appropriate withdrawal times to slaughter should be followed.

Inadequate ventilation, crowding, commingling of animals from various farms (feedlot or sale barn situations), poor nutrition, failure of passive transfer of antibodies, transportation, and other stresses have all been associated with pneumonia outbreaks. Control and prevention lies with correction of the predisposing factors whenever practical. At present, there are no bacterins that have proved effective for control of these pneumonias.

PSEUDOMONAS AERUGINOSA

Pseudomonas aeruginosa is member of the Gamma Proteobacteria class of Bacteria. It is a Gram-negative, aerobic rod belonging to the bacterial family Pseudomonadaceae. Since the revisionist taxonomy based on conserved macromolecules (*e.g.*, 16S ribosomal RNA) the family includes only members of the genus Pseudomonas which are cleaved into eight groups. Pseudomonas aeruginosa is the type species of its group. which contains 12 other members.

Like other members of the genus, *Pseudomonas aeruginosa* is a free-living bacterium, commonly found in soil and water. However, it occurs regularly on the surfaces of plants and occasionally on the surfaces of animals. Members of the genus are well known to plant microbiologists because they are one of the few groups of bacteria that are true pathogens of plants. In fact, *Pseudomonas aeruginosa* is occasionally a pathogen of plants. However, *Pseudomonas aeruginosa* has become increasingly recognised as an emerging opportunistic pathogen of clinical relevance. Several different epidemiological studies track its occurrence as a nosocomial pathogen and indicate that antibiotic resistance is increasing in clinical isolates.

Pseudomonas aeruginosa is an opportunistic pathogen, meaning that it exploits some break in the host defences to initiate an infection. In fact, *Pseudomonas aeruginosa* is the epitome of an opportunistic pathogen of humans. The bacterium almost never infects uncompromised tissues, yet there is hardly any tissue that it cannot infect if the tissue defences are compromised in some manner.

It causes urinary tract infections, respiratory system infections, dermatitis, soft tissue infections, bacteremia, bone and joint infections, gastrointestinal infections and a variety of systemic infections, particularly in patients with severe burns and in cancer and AIDS patients who are immunosuppressed. Pseudomonas aeruginosa infection is a serious problem in patients hospitalised with cancer, cystic fibrosis, and burns. The case fatality rate in these patients is near 50 percent. *Pseudomonas aeruginosa* is primarily a nosocomial pathogen. According to the CDC, the overall incidence of *P. aeruginosa* infections in U.S., hospitals averages about 0.4 percent (4 per 1000 discharges), and the bacterium is the fourth most commonly-isolated nosocomial pathogen accounting for 10.1 percent of all hospital-acquired infections.

CHARACTERISTICS

Pseudomonas aeruginosa is a Gram-negative rod measuring 0.5 to 0.8 μm by 1.5 to 3.0 μm. Almost all strains are motile by means of a single polar flagellum.

The bacterium is ubiquitous in soil and water, and on surfaces in contact with soil or water. Its metabolism is respiratory and never fermentative, but it will grow in the absence of O_2 if NO_3 is available as a respiratory electron acceptor.

The typical *Pseudomonas* bacterium in nature might be found in a biofilm, attached to some surface or substrate, or in a planktonic form, as a unicellular organism, actively swimming by means of its flagellum. *Pseudomonas* is one of the most vigourous, fast-swimming bacteria seen in hay infusions and pond water samples.

In its natural habitat *Pseudomonas aeruginosa* is not particularly distinctive as a pseudomonad, but it does have a combination of physiological traits that are noteworthy and may relate to its pathogenesis.

- *Pseudomonas aeruginosa* has very simple nutritional requirements. It is often observed "growing in distilled water", which is evidence of its minimal nutritional needs. In the laboratory, the simplest medium for growth of *Pseudomonas aeruginosa* consists of acetate as a source of carbon and ammonium sulfate as a source of nitrogen.
- *P. aeruginosa* possesses the metabolic versatility for which pseudomonads are so renowned. Organic growth factors are not required, and it can use more than seventy-five organic compounds for growth.
- Its optimum temperature for growth is 37 degrees, and it is able to grow at temperatures as high as 42 degrees.
- It is tolerant to a wide variety of physical conditions, including temperature. It is resistant to high concentrations of salts and dyes, weak antiseptics, and many commonly used antibiotics.
- *Pseudomonas aeruginosa* has a predilection for growth in moist environments, which is probably a reflection of its natural existence in soil and water.

These natural properties of the bacterium undoubtedly contribute to its ecological success as an opportunistic pathogen. They also help explain the ubiquitous nature of the organism and its prominence as a nosocomial pathogen.

P. aeruginosa isolates may produce three colony types. Natural isolates from soil or water typically produce a small, rough colony. Clinical samples, in general, yield one or another of two smooth colony types. One type has a fried-egg appearance which is large, smooth, with flat edges and an elevated appearance. Another type, frequently obtained from respiratory and urinary tract secretions, has a mucoid appearance, which is attributed to the production of alginate slime. The smooth and mucoid colonies are presumed to play a role in colonisation and virulence.

P. aeruginosa strains produce two types of soluble pigments, the fluorescent pigment pyoverdin and the blue pigment pyocyanin. The latter is produced abundantly in media of low-iron content and functions in iron metabolism in the bacterium. Pyocyanin (from "pyocyaneus") refers to "blue pus", which is a characteristic of suppurative infections caused by *Pseudomonas aeruginosa*.

RESISTANCE TO ANTIBIOTICS

Pseudomonas aeruginosa is notorious for its resistance to antibiotics and is, therefore, a particularly dangerous and dreaded pathogen. The bacterium is naturally resistant to many antibiotics due to the permeabiliity barrier afforded by its Gram-negative outer membrane. Also, its tendency to colonise surfaces in a biofilm form makes the cells impervious to therapeutic concentrations antibiotics. Since its natural habitat is the soil, living in association with the bacilli, actinomycetes and molds, it has developed resistance to a variety of their naturally-occuring antibiotics. Moreover, *Pseudomonas* maintains antibiotic resistance plasmids, both R-factors and RTFs, and it is able to transfer these genes by means of the bacterial mechanisms of horizontal gene transfer (HGT), mainly transduction and conjugation. Only a few antibiotics are effective against *Pseudomonas aeruginosa*, including fluoroquinolones, gentamicin and imipenem, and even these antibiotics are not effective against all strains. The futility of treating *Pseudomonas* infections with antibiotics is most dramatically illustrated in cystic fibrosis patients, virtually all of whom eventually become infected with a strain that is so resistant that it cannot be treated.

DIAGNOSIS

Diagnosis of *P.aeruginosa* infection depends upon isolation and laboratory identification of the bacterium. It grows well on most laboratory media and commonly is isolated on blood agar or eosin-methylthionine blue agar. It is identified on the basis of its Gram morphology, inability to ferment lactose, a positive oxidase reaction, its fruity odour, and its ability to grow at 42°C. Fluorescence under ultraviolet light is helpful in early identification of *P. aeruginosa* colonies. Fluorescence is also used to suggest the presence of *P. aeruginosa* in wounds.

PATHOGENESIS

For an opportunistic pathogen such as *Pseudomonas aeruginosa*, the disease process begins with some alteration or circumvention of normal host defences. The pathogenesis of *Pseudomonas* infections is multifactorial, as suggested by the number and wide array of virulence determinants possessed by the bacterium. Multiple and diverse determinants of virulence are expected in the wide range of diseases caused, which include septicemia, urinary tract infections, pneumonia, chronic lung infections, endocarditis, dermatitis, and osteochondritis.

Most *Pseudomonas* infections are both invasive and toxinogenic.

The ultimate Pseudomonas infection may be seen as composed of three distinct stages:

1. Bacterial attachment and colonisation;
2. Local invasion;
3. Disseminated systemic disease.

However, the disease process may stop at any stage. Particular bacterial determinants of virulence mediate each of these stages and are ultimately responsible for the characteristic syndromes that accompany the disease.

COLONISATION

Although colonisation usually precedes infections by *Pseudomonas aeruginosa*, the exact source and mode of transmission of the pathogen are often unclear because of its ubiquitous presence in the environment. It is sometimes present as part of the normal flora of humans, although the prevalence of colonisation of healthy individuals outside the hospital is relatively low (estimates range from 0 to 24 percent depending on the anatomical locale).

The pili of *Pseudomonas aeruginosa* will adhere to the epithelial cells of the upper respiratory tract and, by inference, to other epithelial cells as well. These adhesins appear to bind to specific galactose or mannose or sialic acid receptors on epithelial cells. Colonisation of the respiratory tract by *Pseudomonas* requires pili adherence and may be aided by production of a protease enzyme that degrades fibronectin in order to expose the underlying pilus receptors on the epithelial cell surface. Tissue injury may also play a role in colonisation of the respiratory tract, since *P. aeruginosa* will adhere to tracheal epithelial cells of mice infected with influenza virus but not to normal tracheal epithelium. This has been called opportunistic adherence, and it may be an important step in *Pseudomonas* keratitis and urinary tract infections, as well as infections of the respiratory tract.

The receptor on tracheal epithelial cells for *Pseudomonas* pili is probably sialic acid (N-acetylneuraminic acid). Mucoid strains, which produce an exopolysaccharide (alginate), have an additional or alternative adhesin which attaches to the tracheobronchial mucin (N-acetylglucosamine). Besides pili and the mucoid polysaccharide, there are possibly other cell surface adhesins utilised by *Pseudomonas* to colonise the respiratory epithelium or mucin. Also, it is possible that surface-bound exoenzyme S could serve as an adhesin for glycolipids on respiratory cells. The mucoid exopolysaccharide produced by *P. aeruginosa* is a repeating polymer of mannuronic and glucuronic acid referred to as alginate. Alginate slime forms the matrix of the *Pseudomonas* biofilm which anchors the cells to their environment and in medical situations, it protects the bacteria from the host defences such as lymphocytes, phagocytes, the ciliary action of the respiratory tract, antibodies and complement. Biofilm mucoid strains of *Pseudomonas* are also less susceptible to antibiotics than their planktonic counterparts. Mucoid strains of *P. aeruginosa* are most often isolated from patients with cystic fibrosis and they are usually found in lung tissues from such individuals.

INVASION

The ability of *Pseudomonas aeruginosa* to invade tissues depends upon production of extracellular enzymes and toxins that break down physical barriers and damage host cells, as well as resistance to phagocytosis and the host immune defences. As mentioned above, the bacterial capsule or slime layer effectively protects cells from opsonisation by antibodies, complement deposition, and phagocyte engulfment.

Two extracellular proteases have been associated with virulence that exert their activity at the invasive stage:

1. Elastase and
2. Alkaline protease.

Elastase has several activities that relate to virulence. The enzyme cleaves collagen, IgG, IgA, and complement. It also lyses fibronectin to expose receptors for bacterial attachment on the mucosa of the lung. Elastase disrupts the respiratory epithelium and interferes with ciliary function. Alkaline protease interferes with fibrin formation and will lyse fibrin. Together, elastase and alkaline protease destroy the ground substance of the cornea and other supporting structures composed of fibrin and elastin. Elastase and alkaline protease together are also reported to cause the inactivation of gamma interferon (IFN) and tumour necrosis factor (TNF).

Pseudomonas aeruginosa produces three other soluble proteins involved in invasion: a cytotoxin (mw 25 kDa) and two hemolysins. The cytotoxin is a pore-forming protein. It was originally named leukocidin because of its effect on neutrophils, but it appears to be cytotoxic for most eucaryotic cells. Of the two hemolysins, one is a phospholipase and the other is a lecithinase. They appear to act synergistically to break down lipids and lecithin. The cytotoxin and hemolysins contribute to invasion through their cytotoxic effects on neutrophils, lymphocytes and other eucaryotic cells.

One *Pseudomonas* pigment is probably a determinant of virulence for the pathogen. The blue pigment, pyocyanin, impairs the normal function of human nasal cilia, disrupts the respiratory epithelium, and exerts a proinflammatory effect on phagocytes. A derivative of pyocyanin, pyochelin, is a siderophore that is produced under low-iron conditions to sequester iron from the environment for growth of the pathogen. It could play a role in invasion if it extracts iron from the host to permit bacterial growth in a relatively iron-limited environment. No role in virulence is known for the fluorescent pigments.

DISSEMINATION

Blood stream invasion and dissemination of *Pseudomonas* from local sites of infection is probably mediated by the same cell-associated and extracellular products responsible for the localised disease, although it is not entirely clear how the bacterium produces systemic illness. *P. aeruginosa* is resistant to phagocytosis and the serum bactericidal response due to its mucoid capsule and possibly LPS. The proteases inactivate complement, cleave IgG antibodies, and inactivate IFN, TNF and probably other cytokines. The Lipid A moiety of *Pseudomonas* LPS (endotoxin) mediates the usual pathologic aspects of Gram-negative septicemia, *e.g.*, fever, hypotension, intravascular coagulation, etc. It is also assumed that *Pseudomonas* Exotoxin A exerts some pathologic activity during the dissemination stage.

TOXINOGENESIS

Pseudomonas aeruginosa produces two extracellular protein toxins, Exoenzyme S and Exotoxin A. Exoenzyme S has the characteristic subunit structure of the A-component of a bacterial toxin, and it has ADP-ribosylating activity (for a variety of eucaryotic proteins) characteristic of many bacterial exotoxins. Exoenzyme S is produced by bacteria growing in burned tissue and may be detected in the blood before the bacteria are. It has led to the suggestion that exoenzyme S may act to impair the function of phagocytic cells in the bloodstream and internal organs as a preparation for invasion by *P. aeruginosa*.

Exotoxin A has exactly the same mechanism of action as the diphtheria toxin; it causes the ADP ribosylation of eucaryotic elongation factor 2 resulting in inhibition of protein synthesis in the affected cell. Although it is partially-identical to diphtheria toxin, it is antigenically-distinct. It utilises a different receptor on host cells than diphtheria toxin, but otherwise it enters cells in the same manner and has the exact enzymatic mechanism. The production of Exotoxin A is regulated by exogenous iron, but the details of the regulatory process are distinctly different in *C. diphtheriae* and *P. aeruginosa*.

Exotoxin A appears to mediate both local and systemic disease processes caused by *Pseudomonas aeruginosa*. It has necrotising activity at the site of bacterial colonisation and is thereby thought to contribute to the colonisation process. Toxinogenic strains cause a more virulent form of pneumonia than non-toxinogenic strains. In terms of its systemic role in virulence, purified Exotoxin A is highly lethal for animals including primates. Indirect evidence involving the role of exotoxin A in disease is seen in the increased chance of survival in patients with *Pseudomonas* septicemia that is correlated with the titer of anti-exotoxin A antibodies in the serum. Also, tox- mutants show a reduced virulence in some models.

Diseases Caused by Pseudomonas Aeruginosa:

- *Endocarditis: Pseudomonas aeruginosa* infects heart valves of IV drug users and prosthetic heart valves. The organism establishes itself on the endocardium by direct invasion from the blood stream.

- *Respiratory Infections*: Respiratory infections caused by *Pseudomonas aeruginosa* occur almost exclusively in individuals with a compromised lower respiratory tract or a compromised systemic defence mechanism. Primary pneumonia occurs in patients with chronic lung disease and congestive heart failure. Bacteremic pneumonia commonly occurs in neutropenic cancer patients undergoing chemotherapy. Lower respiratory tract colonisation of cystic fibrosis patients by mucoid strains of *Pseudomonas aeruginosa* is common and difficult, if not impossible, to eradicate.
- *Bacteremia and Septicemia*: *Pseudomonas aeruginosa* causes bacteremia primarily in immunocompromised patients. Predisposing conditions include hematologic malignancies, immunodeficiency relating to AIDS, neutropenia, diabetes mellitus, and severe burns. Most *Pseudomonas* bacteremia is acquired in hospitals and nursing homes. *Pseudomonas* accounts for about 25 percent of all hospital acquired Gram-negative bacteremias.
- *Central Nervous System Infections: Pseudomonas aeruginosa* causes meningitis and brain abscesses. The organism invades the CNS from a contiguous structure such as the inner ear or paranasal sinus, or is inoculated directly by means of head trauma, surgery or invasive diagnostic procedures, or spreads from a distant site of infection such as the urinary tract.
- *Ear Infections Including External Otitis*: *Pseudomonas aeruginosa* is the predominant bacterial pathogen in some cases of external otitis, including "swimmer's ear". The bacterium is infrequently found in the normal ear, but often inhabits the external auditory canal in association with injury, maceration, inflammation, or simply wet and humid conditions.
- *Eye Infections*: *Pseudomonas aeruginosa* can cause devastating infections in the human eye. It is one of the most common causes of bacterial keratitis, and has been isolated as the etiologic agent of neonatal ophthalmia. *Pseudomonas* can colonise the ocular epithelium by means of a fimbrial attachment to sialic acid receptors. If the defences of the environment are compromised in any way, the bacterium can proliferate rapidly through the production of enzymes such as elastase, alkaline protease and exotoxin A, and cause a rapidly destructive infection that can lead to loss of the entire eye.
- *Bone and Joint Infections*: *Pseudomonas* infections of bones and joints result from direct inoculation of the bacteria or the hematogenous spread of the bacteria from other primary sites of infection. Blood-

borne infections are most often seen in IV drug users and in conjunction with urinary tract or pelvic infections. *Pseudomonas aeruginosa* has a particular tropism for fibrocartilagenous joints of the axial skeleton. *Pseudomonas aeruginosa* causes chronic contiguous osteomyelitis, usually resulting from direct inoculation of bone and is the most common pathogen implicated in osteochondritis after puncture wounds of the foot.

- *Urinary Tract Infections:* Urinary tract infections (UTI) caused by Pseudomonas aeruginosa are usually hospital-acquired and related to urinary tract catheterisation, instrumentation or surgery. *Pseudomonas aeruginosa* is the third leading cause of hospital-acquired UTIs, accounting for about 12 percent of all infections of this type. The bacterium appears to be among the most adherent of common urinary pathogens to the bladder uroepithelium. As in the case of *E. coli*, urinary tract infection can occur via an ascending or descending route. In addition, *Pseudomonas* can invade the bloodstream from the urinary tract, and this is the source of nearly 40 percent of *Pseudomonas* bacteremias.
- *Gastrointestinal Infections*: *Pseudomonas aeruginosa* can produce disease in any part of the gastrointestinal tract from the oropharynx to the rectum. As in other forms of *Pseudomonas* disease, those involving the GI tract occur primarily in immunocompromised individuals. The organism has been implicated in perirectal infections, pediatric diarrhea, typical gastroenteritis, and necrotising enterocolitis. The GI tract is also an important portal of entry in *Pseudomonas* septicemia and bacteremia.
- *Skin and Soft Tissue Infections, Including Wound Infections, Pyoderma and Dermatitis*: *Pseudomonas aeruginosa* can cause a variety of skin infections, both localised and diffuse. The common predisposing factors are breakdown of the integument which may result from burns, trauma or dermatitis; high moisture conditions such as those found in the ear of swimmers and the toe webs of athletes, hikers and combat troops, in the perineal region and under diapers of infants, and on the skin of whirlpool and hot tub users. Individuals with AIDS are easily infected. *Pseudomonas* has also been implicated in folliculitis and unmanageable forms of acne vulgaris.

HOST DEFENCES

Most strains of *P. aeruginosa*are resistant to killing in serum alone, but the addition of polymorphonuclear leukocytes results in bacterial killing. Killing is most efficient in the presence of type-specific opsonising antibodies, directed primarily at the antigenic determinants of LPS. This suggests that phagocytosis is an important defence and that opsonising antibody is the principal functional antibody in protecting from *P. aeruginosa* infections. Once *P. aeruginosa* infection is established, other antibodies, such as antitoxin, may be important in controlling disease.

The observation that patients with diminished antibody responses (caused by underlying disease or associated therapy) have more frequent and more serious *P. aeruginosa* infections underscores the importance of antibody-mediated immunity in controlling *Pseudomonas* infections. unfortunately, cystic fibrosis is the exception. Most cystic fibrosis patients have high levels of circulating antibodies to bacterial antigens, but are unable to clear *P. aeruginosa* efficiently from their lungs. Cell-mediated immunity does not seem to play a major role in resistance or defence against *Pseudomonas* infections.

EPIDEMIOLOGY AND CONTROL OF PSEUDOMONAS AERUGINOSA INFECTIONS

Pseudomonas aeruginosa is a common inhabitant of soil, water, and vegetation. It is found on the skin of some healthy persons and has been isolated from the throat (5 percent) and stool (3 percent) of non-hospitalised patients. In some studies, gastrointestinal carriage rates increased in hospitalised patients to 20 percent within 72 hours of admission. Within the hospital, *P. aeruginosa* finds numerous reservoirs: disinfectants, respiratory equipment, food, sinks, taps, toilets, showers and mops. Furthermore, it is constantly reintroduced into the hospital environment on fruits, plants, vegetables, as well by visitors and patients transferred from other facilities. Spread occurs from patient to patient on the hands of hospital personnel, by direct patient contact with contaminated reservoirs, and by the ingestion of contaminated foods and water.

The spread of *P. aeruginosa* can best be controlled by observing proper isolation procedures, aseptic technique, and careful cleaning and monitoring of respirators, catheters, and other instruments. Topical therapy of burn wounds with antibacterial agents such as silver sulfadiazine, coupled with surgical debridement, dramatically reduces the incidence of *P. aeruginosa* sepsis in burn patients.

Pseudomonas aeruginosa is frequently resistant to many commonly used antibiotics. Although many strains are susceptible to gentamicin, tobramycin, colistin, and fluoroquinolins, resistant forms have developed. The combination of gentamicin and carbenicillin is frequently used to treat severe *Pseudomonas* infections. Several types of vaccines are being tested, but none is currently available for general use.

MYCOBACTERIUM OF BACTERIA

Mycobacterium is a genus of Actinobacteria, given its own family, the

Mycobacteriaceae. The genus includes pathogens known to cause serious diseases in mammals, including tuberculosis (*Mycobacterium tuberculosis*) and leprosy (*Mycobacterium leprae*). The Greek prefix *myco* means *fungus*, alluding to the way mycobacteria have been observed to grow in a mold-like fashion on the surface of liquids when cultured.

ICROBIOLOGIC CHARACTERISTICS

Mycobacteria are aerobic and non-motile bacteria (except for the species *Mycobacterium marinum*, which has been shown to be motile within macrophages) that are characteristically acid-alcohol-fast. Mycobacteria do not contain endospores or capsules and are usually considered Gram-positive. *Mycobacterium marinum* and perhaps *M. bovis* have been shown to sporulate; however, this has been contested by further research. While mycobacteria do not seem to fit the Gram-positive category from an empirical standpoint (*i.e.*, in general, they do not retain the crystal violet stain well), they are classified as an acid-fast Gram-positive bacterium due to their lack of an outer cell membrane. All *Mycobacterium* species share a characteristic cell wall, thicker than in many other bacteria, which is hydrophobic, waxy, and rich in mycolic acids/mycolates. The cell wall consists of the hydrophobic mycolate layer and a peptidoglycan layer held together by a polysaccharide, arabinogalactan. The cell wall makes a substantial contribution to the hardiness of this genus. The biosynthetic pathways of cell wall components are potential targets for new drugs for tuberculosis.

Many *Mycobacterium* species adapt readily to growth on very simple substrates, using ammonia or amino acids as nitrogen sources and glycerol as a carbon source in the presence of mineral salts. Optimum growth temperatures vary widely according to the species and range from 25 °C to over 50 °C.

Some species can be very difficult to culture (*i.e.*, they are fastidious), sometimes taking over two years to develop in culture. Further, some species also have extremely long reproductive cycles — *M. leprae*, may take more than 20 days to proceed through one division cycle (for comparison, some *E. coli* strains take only 20 minutes), making laboratory culture a slow process. In addition, the availability of genetic manipulation techniques still lags far behind that of other bacterial species.

A natural division occurs between slowly– and rapidly–growing species. Mycobacteria that form colonies clearly visible to the naked eye within seven days on subculture are termed rapid growers, while those requiring longer periods are termed slow growers. Mycobacteria cells are straight or slightly curved rods between 0.2 and 0.6 μm wide by 1.0 and 10 μm long.

PIGMENTATION

Some mycobacteria produce carotenoid pigments without light. Others require photoactivation for pigment production.

- Photochromogens (Group I): Produce non-pigmented colonies when grown in the dark and pigmented colonies only after exposure to light and reincubation.
  - Ex: *M. kansasii, M. marinum, M. simiae.*
- Scotochromogens (Group II): Produce deep yellow to orange colonies when grown in the presence of either the light or the dark.
  - Ex: *M. scrofulaceum, M. gordonae, M. xenopi, M. szulgai.*
- Non-chromogens (Groups III & IV): Non-pigmented in the light and dark or have only a pale yellow, buff or tan pigment that does not intensify after light exposure.
  - Ex: *M. tuberculosis, M. avium-intra-cellulare, M. bovis, M. ulcerans*
  - Ex: *M. fortuitum, M. chelonae*

STAINING CHARACTERISTICS

Mycobacteria are classical acid-fast organisms. Stains used in evaluation of tissue specimens or microbiological specimens include Fite's stain, Ziehl-Neelsen stain, and Kinyoun stain.

Mycobacteria appear phenotypically most closely related to members of *Nocardia*, *Rhodococcus* and *Corynebacterium*.

ECOLOGICAL CHARACTERISTICS

Mycobacteria are widespread organisms, typically living in water (including tap water treated with chlorine) and food sources. Some, however, including the tuberculosis and the leprosy organisms, appear to be obligate parasites and are not found as free-living members of the genus.

PATHOGENICITY

Mycobacteria can colonise their hosts without the hosts showing any adverse signs. For example, billions of people around the world have asymptomatic infections of *M. tuberculosis*.

Mycobacterial infections are notoriously difficult to treat. The organisms are hardy due to their cell wall, which is neither truly Gram negative nor positive. In addition, they are naturally resistant to a number of antibiotics that disrupt cell-wall biosynthesis, such as penicillin. Due to their unique cell wall, they can survive long exposure to acids, alkalis, detergents, oxidative bursts, lysis by complement, and many antibiotics. Most mycobacteria are susceptible to the antibiotics clarithromycin and rifamycin, but antibiotic-resistant strains have emerged.

As with other bacterial pathogens, surface and secreted proteins of *M. tuberculosis* contribute significantly to the virulence of this organism. There is an increasing list of extracytoplasmic proteins proven to have a function in the virulence of *M. tuberculosis*.

MEDICAL CLASSIFICATION

Mycobacteria can be classified into several major groups for purpose of diagnosis and treatment: *M. tuberculosis* complex, which can cause tuberculosis: *M. tuberculosis*, *M. bovis*, *M. africanum*, and *M. microti*; *M. leprae*, which causes Hansen's disease or leprosy; Non-tuberculous mycobacteria (NTM) are all the other mycobacteria, which can cause pulmonary disease resembling tuberculosis, lymphadenitis, skin disease, or disseminated disease.

MYCOSIDES

Mycosides are phenolic alcohols (such as phenolphthiocerol) that were shown to be components of *Mycobacterium* glycolipids that are termed glycosides of phenolphthiocerol dimycocerosate. There are 18 and 20 carbon atoms in mycosides A, and B, respectively.

GENOMICS

Comparative analyses of mycobacterial genomes have identified several conserved indels and signature proteins that are uniquely found in all sequenced species from the genus *Mycobacterium*. Additionally, 14 proteins are found only in the species from the genera *Mycobacterium* and *Nocardia*, suggesting that these two genera are closely related.

SPECIES

Phylogenetic Position of the Tubercle Bacilli within the Genus *Mycobacterium* The blue triangle corresponds to tubercle bacilli sequences that are identical or differing by a single nucleotide. The sequences of the genus *Mycobacterium* that matched most closely to those of *M. tuberculosis* were retrieved from the BIBI database and aligned with those obtained for 17 smooth and MTBC strains. The unrooted neighbour-joining tree is based on 1,325 aligned nucleotide positions of the 16S rRNA gene. The scale gives the pairwise distances after Jukes-Cantor correction. Bootstrap support values higher than 90 per cent are indicated at the nodes. Phenotypic tests can be used to identify and distinguish different Mycobacteria species and strains. In older systems, mycobacteria are grouped based upon their appearance and rate of growth. However, these are symplesiomorphies, and more recent classification is based upon cladistics.

SLOWLY GROWING

Mycobacterium tuberculosis complex:

- *Mycobacterium tuberculosis* complex (MTBC) members are causative agents of human and animal tuberculosis. Species in this complex include:
  - *M. tuberculosis*, the major cause of human tuberculosis

  - *M. bovis*
  - *M. bovis BCG*
  - *M. africanum*
  - *M. canetti*
  - *M. caprae*
  - *M. microti*
  - *M. pinnipedii*

Mycobacterium avium complex:

- *Mycobacterium avium* complex (MAC) is a group of species that, in a disseminated infection but not lung infection, used to be a significant cause of death in AIDS patients. Species in this complex include:
  - *M. avium*
  - *M. avium paratuberculosis*, which has been implicated in Crohn's disease in humans and Johne's disease in cattle and sheep
  - *M. avium silvaticum*
  - *M. avium "hominissuis"*
  - *M. colombiense*
  - *M. indicus pranii*

Mycobacterium gordonae clade:

- *M. asiaticum*
- *M. gordonae*

Mycobacterium kansasii clade:

- *M. gastri*
- *M. kansasii*

Mycobacterium Non-chromogenicum/Terrae Clade:

- *M. hiberniae*
- *M. non-chromogenicum*
- *M. terrae*
- *M. triviale*

Mycolactone-producing mycobacteria:

- *M. ulcerans*, which causes the "Buruli", or "Bairnsdale, ulcer"
- *M. pseudoshottsii*
- *M. shottsii*

Mycobacterium simiae clade:

- *M. triplex*
- *M. genavense*
- *M. florentinum*
- *M. lentiflavum*
- *M. palustre*
- *M. kubicae*
- *M. parascrofulaceum*
- *M. heidelbergense*

- *M. interjectum*
- *M. simiae*

Ungrouped:

- *M. branderi*
- *M. cookii*
- *M. celatum*
- *M. bohemicum*
- *M. haemophilum*
- *M. malmoense*
- *M. szulgai*
- *M. leprae*, which causes leprosy
- *M. lepraemurium*
- *M. lepromatosis*, another (less significant) cause of leprosy, described in 2008
- *M. africanum*
- *M. botniense*
- *M. chimaera*
- *M. conspicuum*
- *M. doricum*
- *M. farcinogenes*
- *M. heckeshornense*
- *M. intracellulare*
- *M. lacus*
- *M. marinum*
- *M. monacense*
- *M. montefiorense*
- *M. murale*
- *M. nebraskense*
- *M. saskatchewanense*
- *M. scrofulaceum*
- *M. shimoidei*
- *M. tusciae*
- *M. xenopi*

INTERMEDIATE GROWTH RATE

- *M. intermedium*

RAPIDLY GROWING

Mycobacterium chelonae clade:

- *M. abscessus*
- *M. chelonae*
- *M. bolletii*

Mycobacterium fortuitum clade:

- *M. fortuitum*
- *M. fortuitum subsp. acetamidolyticum*
- *M. boenickei*
- *M. peregrinum*
- *M. porcinum*
- *M. senegalense*
- *M. septicum*
- *M. neworleansense*
- *M. houstonense*
- *M. mucogenicum*
- *M. mageritense*
- *M. brisbanense*
- *M. cosmeticum*

Mycobacterium parafortuitum clade:

- *M. parafortuitum*
- *M. austroafricanum*
- *M. diernhoferi*
- *M. hodleri*
- *M. neo-aurum*
- *M. frederiksbergense*

Mycobacterium vaccae clade:

- *M. aurum*
- *M. vaccae*

CF:

- *M. chitae*
- *M. fallax*

Ungrouped:

- *M. confluentis*
- *M. flavescens*
- *M. madagascariense*
- *M. phlei*
- *M. smegmatis*
  - *M. goodii*
  - *M. wolinskyi*
- *M. thermoresistibile*
- *M. gadium*
- *M. komossense*
- *M. obuense*
- *M. sphagni*
- *M. agri*
- *M. aichiense*

- *M. alvei*
- *M. arupense*
- *M. brumae*
- *M. canariasense*
- *M. chubuense*
- *M. conceptionense*
- *M. duvalii*
- *M. elephantis*
- *M. gilvum*
- *M. hassiacum*
- *M. holsaticum*
- *M. immunogenum*
- *M. massiliense*
- *M. moriokaense*
- *M. psychrotolerans*
- *M. pyrenivorans*
- *M. vanbaalenii*
- *M. pulveris*

Ungrouped:

- *M. arosiense*
- *M. aubagnense*
- *M. caprae*
- *M. chlorophenolicum*
- *M. fluoroanthenivorans*
- *M. kumamotonense*
- *M. novocastrense*
- *M. parmense*
- *M. phocaicum*
- *M. poriferae*
- *M. rhodesiae*
- *M. seoulense*
- *M. tokaiense*

MYCOBACTERIOPHAGE

Mycobacteria can be infected by Mycobacteriophage, bacterial viruses that may be used in the future to treat tuberculosis and related diseases by phage therapy.

MYCOBACTERIUM TUBERCULOSIS

Mycobacterium tuberculosis (MTB) is a pathogenic bacterial species in the family Mycobacteriaceae and the causative agent of most cases of tuberculosis (TB). First discovered in 1882 by Robert Koch, *M. tuberculosis* has an unusual, waxy coating on its cell surface (primarily mycolic acid), which makes the cells impervious to Gram staining. Acid-fast detection techniques are used instead. The physiology of *M. tuberculosis* is highly aerobic and requires high levels of oxygen. Primarily a pathogen of the mammalian respiratory system, MTB infects the lungs. The most frequently used diagnostic methods for TB are the tuberculin skin test, acid-fast stain, and chest radiographs. The *M. tuberculosis* genome was sequenced in 1998.

PATHOPHYSIOLOGY

M. tuberculosis requires oxygen to grow. It does not retain any bacteriological stain due to high lipid content in its wall, hence Ziehl-Neelsen staining, or acid-fast staining, is used. Despite this, it is gram-positive bacteria. While mycobacteria do not seem to fit the Gram-positive category from an empirical standpoint (*i.e.*, they do not retain the crystal violet stain), they are classified as acid-fast Gram-positive bacteria due to their lack of an outer cell membrane.

M. tuberculosis divides every 15–20 hours, which is extremely slow compared to other bacteria, which tend to have division times measured in minutes (*Escherichia coli* can divide roughly every 20 minutes). It is a small bacillus that can withstand weak disinfectants and can survive in a dry state for weeks. Its unusual cell wall, rich in lipids (*e.g.*, mycolic acid), is likely responsible for this resistance and is a key virulence factor.

When in the lungs, *M. tuberculosis* is taken up by alveolar macrophages, but they are unable to digest the bacterium. Its cell wall prevents the fusion of the phagosome with a lysosome. Specifically, *M. tuberculosis* blocks the bridging molecule, early endosomal autoantigen 1 (EEA1); however, this blockade does not prevent fusion of vesicles filled with nutrients. Consequently, the bacteria multiply unchecked within the macrophage. The bacteria also carried the *UreC* gene, which prevents acidification of the phagosome. The bacteria also evade macrophage-killing by neutralising reactive nitrogen intermediates.

The ability to construct *M. tuberculosis* mutants and test individual gene products for specific functions has significantly advanced our understanding of the pathogenesis and virulence factors of *M. tuberculosis*. Many secreted and exported proteins are known to be important in pathogenesis.

STRAIN VARIATION

M. tuberculosis comes from the genus *Mycobacterium*, which is composed of approximately 100 recognised and proposed species. The most familiar of the species are *M. tuberculosis* and *M. leprae* (leprosy).

M. tuberculosis is genetically diverse, which results in significant phenotypic differences between clinical isolates. Different strains of *M. tuberculosis* are associated with different geographic regions. However, phenotypic studies suggest that strain variation never has implications for the development of new diagnostics and vaccines. Microevolutionary variation does affect the relative fitness and transmission dynamics of antibiotic-resistant strains. Typing of strains is useful in the investigation of tuberculosis outbreaks, because it gives the investigator evidence for-or-against transmission from person to person. Consider the situation where person A has tuberculosis and believes that he acquired it from person B. If the bacteria isolated from each person belong to different types, then transmission from B to A is definitively disproved; on the other hand, if the bacteria are the same strain, then this supports (but does not definitively prove) the theory that B infected A. Until the early 2000s, *M. tuberculosis* strains were typed by pulsed field gel electrophoresis (PFGE). This has now been superseded by variable numbers of tandem repeats (VNTR), which is technically easier to perform and allows better discrimination between strains. This method makes use of the presence of repeated DNA sequences within the *M. tuberculosis* genome.

There are three generations of VNTR typing for *M. tuberculosis*. The first scheme, called ETR (exact tandem repeat), used only five loci, but the resolution afforded by these five loci was not as good as PFGE. The second scheme, called MIRU (mycobacterial interspersed repetitive unit) had discrimination as good as PFGE. The third generation (MIRU2) added a further nine loci to bring the total to 24. This provides a degree of resolution greater than PFGE and is currently the standard for typing *M. tuberculosis*.

HYPERVIRULENT STRAINS

Mycobacterium outbreaks are often caused by hypervirulent strains of *M. tuberculosis*. In laboratory experiments, these clinical isolates elicit unusual immunopathology, and may be either hyperinflammatory or hypoinflammatory. Studies have shown the majority of hypervirulent mutants have deletions in their cell wall-modifying enzymes or regulators that respond to environmental stimuli. Studies of these mutants have indicated the mechanisms that enable *M. tuberculosis* to mask its full pathogenic potential, inducing a granuloma that provides a protective niche, and enable the bacilli to sustain a long-term, persistent infection.

MICROSCOPY

M. tuberculosis is characterised by caseating granulomas containing Langhans giant cells, which have a "horseshoe" pattern of nuclei. Organisms are identified by their red colour on acid-fast staining.

GENOME

The genome of the H37Rv strain was published in 1998. Its size is 4 million base pairs, with 3959 genes; 40 per cent of these genes have had their function characterised, with possible function postulated for another 44 per cent. Within the genome are also six pseudogenes.

The genome contains 250 genes involved in fatty acid metabolism, with 39 of these involved in the polyketide metabolism generating the waxy coat. Such large numbers of conserved genes show the evolutionary importance of the waxy coat to pathogen survival.

About 10 per cent of the coding capacity is taken up by the PE/PPE gene families that encode acidic, glycine-rich proteins. These proteins have a conserved N-terminal motif, deletion of which impairs growth in macrophages and granulomas.

Nine non-coding sRNAs have been characterised in *M. tuberculosis*, with a further 56 predicted in a bioinformatics screen.

EVOLUTION

The *Mycobacterium tuberculosis* complex evolved in Africa and most probably in the Horn of Africa. The *M. tuberculosis* group has a number of members that include *Mycobacterium africanum*, *Mycobacterium bovis* (Dassie's bacillus), *Mycobacterium caprae*, *Mycobacterium microti*, *Mycobacterium mungi*, *Mycobacterium orygis* and *Mycobacterium pinnipedii*. This group may also include the *Mycobacterium canettii* clade.

The *M. canettii* clade—which includes *Mycobacterium prototuberculosis*—are a group of smooth colony *Mycobacterium* species. Unlike the established members of the *M. tuberculosis* group they undergo recombination with other species. The majority of the known strains of this group have been isolated from the Horn of Africa.

The established members of the *M. tuberculosis* complex are all clonal in their spread. The main human infecting species have been classified into seven oligotypes: type 1 contains the East African-Indian (EAI) and some Manu (Indian) strains; type 2 is the Beijing group; type 3 consists of the Central Asian (CAS) strains; type 4 of the Ghana and Haarlem (H/T), Latin America-Mediterranean (LAM) and X strains; types 5 and 6 correspond to *Mycobacterium africanum* and are observed predominantly and at very high frequency in West Africa. A seventh type has been isolated from the Horn of Africa. The other species of this complex belong to a number of oligotypes and do not normally infect humans. Type 2 and 3 are more closely related to each other than to the other types. Types 5 and 6 are most closely aligned with the species that do not normally infect humans. Type 3 has been divided into two clades: CAS-Kili (found in Tanzania) and CAS-Delhi (found in India and Saudi Arabia).

The most recent common ancestor of the *M. tuberculosis* complex evolved ~40,000 years ago. The most recent common ancestor of the EAI and LAM strains has been estimated to be 13,700 and 7,000 years ago respectively. The Beijing- CAS strains diverged ~17,100 years ago. All types of the *M. tuberculosis* began their current expansion ~5000 years ago—a period that coincides with the appearance of *Mycobacterium bovis*. The Beijing strain appears to have been the most successful with a ~500 increase in effective population size (N_e) since its expansion began. The least successful of the main linages appears to be have been those limited to Africa where they have undergone an N_e increase of only 5 fold. Since its initial evolution *M. bovis* has undergone an expansion of its N_e of ~65 fold.

HISTORY

M. tuberculosis, then known as the "tubercle bacillus", was first described on 24 March 1882 by Robert Koch, who subsequently received the Nobel Prize in physiology or medicine for this discovery in 1905; the bacterium is also known as "Koch's bacillus". Tuberculosis has existed throughout history, but the name has changed frequently over time. In 1720, though, the history of tuberculosis started to take shape into what is known of it today; as the physician Benjamin Marten described in his *A Theory of Consumption*, tuberculosis may be caused by small living creatures that are transmitted through the air to other patients.

ENTEROBACTERIACEAE FAMILY

The Enterobacteriaceae is a large family of Gram-negative bacteria that includes, along with many harmless symbionts, many of the more familiar pathogens, such as Salmonella, Escherichia coli, Yersinia pestis, Klebsiella and Shigella. Other disease-causing bacteria in this family include Proteus, Enterobacter, Serratia, and Citrobacter. This family is the only representative in the order Enterobacteriales of the class Gammaproteobacteria in the phylum Proteobacteria. Phylogenetically, in the Enterobacteriales, several peptidoglycan-less insect endosymbionts form a sister clade to the Enterobacteriaceae, but as they are not validly described, this group is not officially a taxon; examples of these species are Sodalis, Buchnera, Wigglesworthia, Baumannia and Blochmannia, but not formers rickettsias. Members of the Enterobacteriaceae can be trivially referred to as enterobacteria, as several members live in the intestines of animals. In fact, the etymology of the family is enterobacterium with the suffix to designate a family (aceae) — not after the genus Enterobacter (which would be "Enterobacteraceae")— and the type genus is Escherichia.

CHARACTERISTICS

Members of the Enterobacteriaceae are rod-shaped, and are typically 1-5 ìm in length. Like other proteobacteria, enterobacteria have Gram-negative stains, and they are facultative anaerobes, fermenting sugars to produce lactic acid and various other end products. Most also reduce nitrate to nitrite, although exceptions exist (*e.g. Photorhabdus*). Unlike most similar bacteria, enterobacteria generally lack cytochrome C oxidase, although there are exceptions (*e.g. Plesiomonas shigelloides*). Most have many flagella used to move about, but a few genera are non-motile. They are not spore-forming. Catalase reactions vary among Enterobacteriaceae. Many members of this family are a normal part of the gut flora found in the intestines of humans and other animals, while others are found in water or soil, or are parasites on a variety of different animals and plants. *Escherichia coli* is one of the most important model organisms, and its genetics and biochemistry have been closely studied. Most members of Enterobacteriaceae have peritrichous, type I fimbriae involved in the adhesion of the bacterial cells to their hosts. Some enterobacteria produce endotoxins. Endotoxins reside in the cell cytoplasm and are released when the cell dies and the cell wall disintegrates. Some members of the Enterobacteriaceae family produce a systemic infection into the blood stream when all the dead bacterial cells release their endotoxins. This is known as endotoxic shock, and can be rapidly fatal.

IDENTIFICATION

To identify different genera of Enterobacteriaceae, a microbiologist may run a series of tests in the lab, these include:

- Phenol red
- Tryptone broth
- Phenylalanine agar for detection of production of deaminase, which converts phenylalanine to phenylpyruvic acid
- Methyl red or Voges-Proskauer tests depend on the digestion of glucose. The methyl red tests for acid endproducts. The Voges Proskauer tests for the production of acetylmethylcarbinol.
- Catalase test on nutrient agar tests for the production of catalase enzyme, which splits hydrogen peroxide and releases oxygen gas.
- Oxidase test on nutrient agar tests for the production of the enzyme oxidase, which reacts with an aromatic amine to produce a purple colour.
- Nutrient gelatin tests to detect activity of the enzyme gelatinase.

In a clinical setting, three species make up 80 to 95 per cent of all isolates identified. These are *Escherichia coli*, *Klebsiella pneumoniae* and *Proteus mirabilis*.

ANTIBIOTIC RESISTANCE

Several Enterobacteriacea strains have been isolated which are resistant to antibiotics including carbapenem, which are often claimed as "the last line of antibiotic defence" against resistant organisms. For instance, some *Klebsiella pneumonia* strains are carbapenem resistant.

EXAMPLES/CLASSIFICATION

The following inexhaustive list details bacterial genera classified as members of Enterobacteriaceae.

GENERA

- *Alishewanella*
- *Alterococcus*
- *Aquamonas*
- *Aranicola*
- *Arsenophonus*
- *Azotivirga*
- *Blochmannia*
- *Brenneria*
- *Buchnera*
- *Budvicia*
- *Buttiauxella*
- *Cedecea*
- *Citrobacter*
- *Cronobacter*
- *Dickeya*
- *Edwardsiella*
- *Enterobacter*
- *Erwinia, e.g. Erwinia amylovora, Erwinia tracheiphila, Erwinia carotovora*, etc.
- *Escherichia, e.g. Escherichia coli*
- *Ewingella*
- *Grimontella*
- *Hafnia*
- *Klebsiella, e.g. Klebsiella pneumoniae*
- *Kluyvera*
- *Leclercia*
- *Leminorella*
- *Moellerella*
- *Morganella*
- *Obesumbacterium*
- *Pantoea*
- *Pectobacterium*
- *Candidatus* Phlomobacter
- *Photorhabdus, e.g. Photorhabdus luminescens*
- *Plesiomonas, e.g. Plesiomonas shigelloides*
- *Pragia*
- *Proteus, e.g. Proteus vulgaris*

- *Providencia*
- *Rahnella*
- *Raoultella*
- *Salmonella*
- *Samsonia*
- *Serratia*, e.g. *Serratia marcescens*
- *Shigella*
- *Sodalis*
- *Tatumella*
- *Trabulsiella*
- *Wigglesworthia*
- *Xenorhabdus*
- *Yersinia*, e.g. *Yersinia pestis*
- *Yokenella*

GENUS OF BACTERIA IN CAMPYLOBACTER

Campylobacter (meaning "twisted bacteria") is a genus of bacteria that are Gram-negative, spiral, and microaerophilic. Motile, with either unipolar or bipolar flagella, the organisms have a characteristic spiral/corkscrew appearance and are oxidase-positive. *Campylobacter jejuni* is now recognised as one of the main causes of bacterial foodborne disease in many developed countries. At least a dozen species of *Campylobacter* have been implicated in human disease, with *C. jejuni* and *C. coli* the most common. *C. fetus* is a cause of spontaneous abortions in cattle and sheep, as well as an opportunistic pathogen in humans.

GENOME

The genomes of several *Campylobacter* species have been sequenced, providing insights into their mechanisms of pathogenesis. The first *Campylobacter* genome to be sequenced was *C. jejuni*, in 2000. *Campylobacter* species contain two flagellin genes in tandem for motility, *flaA* and *flaB*. These genes undergo intergenic recombination, further contributing to their virulence. Non-motile mutants do not colonise.

MOLECULAR SIGNATURES

Comparative genomic analysis has led to the identification of 15 proteins which are uniquely found in members of the genus *Campylobacter* and serve as molecular markers for the genus. Eighteen other proteins were also found which were present in all species except *Campylobacter fetus*, which is the deepest branching *Campylobacter* species. A conserved insertion has also been identified which is present in all *Campylobacter* species except *C. fetus*. Additionally, 28 proteins have been identified present only in *Campylobacter jejuni* and *Campylobacter coli*, indicating a close relationship between these two species.

Five other proteins have also been identified which are only found in *C. jejuni* and serve as molecular markers for the species.

CAUSE

The sites of tissue injury include the jejunum, the ileum, and the colon. Most strains of *C jejuni* produce a toxin (cytolethal distending toxin) that hinders the cells from dividing and activating the immune system. This helps the bacteria to evade the immune system and survive for a limited time in the cells. A cholera-like enterotoxin was once thought to be also made, but this appears not to be the case. The organism produces diffuse, bloody, edematous, and exudative enteritis. Although rarely has the infection been considered a cause of hemolytic uremic syndrome and thrombotic thrombocytopenic purpura, no unequivocal case reports exist. In some cases, a *Campylobacter* infection can be the underlying cause of Guillain–Barré syndrome. Gastrointestinal perforation is a rare complication of ileal infection.

TREATMENT

Diagnosis of the illness is made by testing a specimen of faeces (bowel motion):

- Standard treatment is now azithromycin. Quinolone antibiotics such as ciprofloxacin or levofloxacin are no longer as effective due to resistance.
- Dehydrated children may require intravenous (by vein) fluid treatment in a hospital.
- The illness is contagious, and children must be kept at home until they have been clear of symptoms for at least two days.
- Good hygiene is important to avoid contracting the illness or spreading it to others.
- Intestinal perforation is very rare; increased abdominal pain and collapse require immediate medical attention.

KNOWN RISKS

In January 2013, the UK's Food Standards Agency warned that two-thirds of all raw chicken bought from UK shops was contaminated with campylobacter, affecting an estimated half a million people annually and killing approximately 100.

HISTORY

The symptoms of *Campylobacter* infections were described in 1886 in infants by Theodor Escherich. These infections were named cholera infantum, or summer complaint. The genus was first discovered in 1963; however, the organism was not isolated until 1972. Infections [in the United States] from campylobacter — which is linked to many foods, including poultry, raw milk and produce – has risen up to 14 percent in 2012 compared to 2006-2008. They were at their highest level since 2000.

INFLAMMATORY DISEASES IN PERIODONTITIS

Periodontitis is a set of inflammatory diseases affecting the periodontium, *i.e.*, the tissues that surround and support the teeth. Periodontitis involves progressive loss of the alveolar bone around the teeth, and if left untreated, can lead to the loosening and subsequent loss of teeth. Periodontitis is caused by microorganisms that adhere to and grow on the tooth's surfaces, along with an overly aggressive immune response against these microorganisms. A diagnosis of periodontitis is established by inspecting the soft gum tissues around the teeth with a probe (*i.e.*, a clinical examination) and by evaluating the patient's X-ray films (*i.e.*, a radiographic examination), to determine the amount of bone loss around the teeth. Specialists in the treatment of periodontitis are periodontists; their field is known as "periodontology" or "periodontics".

The word "periodontitis" comes from the Greek *peri*, "around", *odous* (genitive *odontos*), "tooth", and the suffix *-itis*, in medical terminology "inflammation".

CLASSIFICATION

The 1999 classification system for periodontal diseases and conditions listed eight major categories of periodontal diseases, of which 2-6 are termed *destructive* periodontal disease because the damage is essentially irreversible.

The some categories are as follows:

- Gingivitis
- Chronic periodontitis
- Aggressive periodontitis
- Periodontitis as a manifestation of systemic disease
- Necrotising ulcerative gingivitis/periodontitis
- Abscesses of the periodontium
- Combined periodontic-endodontic lesions

Moreover, terminology expressing both the extent and severity of periodontal diseases are appended to the terms above to denote the specific diagnosis of a particular patient or group of patients.

EXTENT

The 'extent' of disease refers to the proportion of the dentition affected by the disease in terms of percentage of sites.

Sites are defined as the positions at which probing measurements are taken around each tooth and, generally, six probing sites around each tooth are recorded, as follows:

1. Mesiobuccal
2. Midbuccal
3. Distobuccal
4. Mesiolingual
5. Midlingual
6. Distolingual

If up to 30 per cent of sites in the mouth are affected, the manifestation is classification as 'localised'; for more than 30 per cent, the term 'generalised' is used.

SEVERITY

The 'severity' of disease refers to the amount of periodontal ligament fibres that have been lost, termed 'clinical attachment loss'.

According to the American Academy of Periodontology, the classification of severity is as follows:

- *Mild:* 1–2 mm (0.039–0.079 in) of attachment loss.
- *Moderate:* 3–4 mm (0.12–0.16 in) of attachment loss.
- *Severe:* e" 5 mm (0.20 in) of attachment loss.

SIGNS AND SYMPTOMS

In the early stages, periodontitis has very few symptoms; and in many individuals the disease has progressed significantly before they seek treatment.

Symptoms may include:

- Redness or bleeding of gums while brushing teeth, using dental floss or biting into hard food (*e.g.*, apples) (though this may occur even in gingivitis, where there is no attachment loss)
- Gum swelling that recurs
- Spitting out blood after brushing teeth
- Halitosis, or bad breath, and a persistent metallic taste in the mouth
- Gingival recession, resulting in apparent lengthening of teeth. (This may also be caused by heavy-handed brushing or with a stiff tooth brush.)
- Deep pockets between the teeth and the gums (pockets are sites where the attachment has been gradually destroyed by collagen-destroying enzymes, known as collagenases)
- Loose teeth, in the later stages (though this may occur for other reasons, as well)

Patients should realise gingival inflammation and bone destruction are largely painless.

Hence, people may wrongly assume painless bleeding after teeth cleaning is insignificant, although this may be a symptom of progressing periodontitis in that patient.

EFFECTS OUTSIDE THE MOUTH

Periodontitis has been linked to increased inflammation in the body, such as indicated by raised levels of C-reactive protein and interleukin-6. It is linked through this to increased risk of stroke, myocardial infarction, and atherosclerosis. It also linked in those over 60 years of age to impairments in delayed memory and calculation abilities. Individuals with impaired fasting glucose and diabetes mellitus have higher degrees of periodontal inflammation, and often have difficulties with balancing their blood glucose level owing to the constant systemic inflammatory state, caused by the periodontal inflammation. Although no causative connection was proved yet, a recent study revealed an epidemiological association between chronic periodontitis and erectile dysfunction.

CAUSES

Periodontitis is an inflammation of the periodontium, *i.e.*, the tissues that support the teeth.

The periodontium consists of four tissues:

- Gingiva, or gum tissue,
- Cementum, or outer layer of the roots of teeth,
- Alveolar bone, or the bony sockets into which the teeth are anchored, and
- Periodontal ligaments (PDLs), which are the connective tissue fibres that run between the cementum and the alveolar bone.

This X-ray film displays two lone-standing mandibular teeth, the lower left first premolar and canine, exhibiting severe bone loss of 30–50 per cent. Widening of the periodontal ligament surrounding the premolar is due to secondary occlusal trauma.

The primary etiology (cause) of gingivitis is poor or ineffective oral hygiene. which leads to the accumulation of a mycotic and bacterial matrix at the gum line, called dental plaque. Other contributors are poor nutrition and underlying medical issues such as diabetes. Diabetics must be meticulous with their homecare to control periodontal disease. New finger nick tests have been approved by the Food and Drug Administration in the US, and are being used in dental offices to identify and screen patients for possible contributory causes of gum disease, such as diabetes.

In some people, gingivitis progresses to periodontitis -- with the destruction of the gingival fibres, the gum tissues separate from the tooth and deepened sulcus, called a periodontal pocket. Subgingival micro-organisms (those that exist under the gum line) colonise the periodontal pockets and cause further inflammation in the gum tissues and progressive bone loss. Examples of secondary etiology are those things that, by definition, cause microbic plaque accumulation, such as restoration overhangs and root proximity.

The excess restorative material that exceeds the natural contours of restored teeth, such as these, are termed "overhangs", and serve to trap microbic plaque, potentially leading to localised periodontitis.

Smoking is another factor that increases the occurrence of periodontitis, directly or indirectly, and may interfere with or adversely affect its treatment.

Ehlers–Danlos syndrome is a periodontitis risk factor and so is the Papillon-Lefèvre syndrome also known as palmoplantar keratoderma.

If left undisturbed, microbial plaque calcifies to form calculus, which is commonly called tartar. Calculus above and below the gum line must be removed completely by the dental hygienist or dentist to treat gingivitis and periodontitis. Although the primary cause of both gingivitis and periodontitis is the microbial plaque that adheres to the tooth surfaces, there are many other modifying factors. A very strong risk factor is one's genetic susceptibility. Several conditions and diseases, including Down syndrome, diabetes, and other diseases that affect one's resistance to infection, also increase susceptibility to periodontitis. Another factor that makes periodontitis a difficult disease to study is that human host response can also affect the alveolar bone resorption. Host response to the bacterial-mycotic insult is mainly determined by genetics; however, immune development may play some role in susceptibility.

According to some researchers periodontitis may be associated with higher stress.

PREVENTION

Daily oral hygiene measures to prevent periodontal disease include:

- Brushing properly on a regular basis (at least twice daily), with the patient attempting to direct the toothbrush bristles underneath the gum-line, helps disrupt the bacterial-mycotic growth and formation of subgingival plaque.
- Flossing daily and using interdental brushes (if the space between teeth is large enough), as well as cleaning behind the last tooth, the third molar, in each quarter
- Using an antiseptic mouthwash: Chlorhexidine gluconate-based mouthwash in combination with careful oral hygiene may cure gingivitis, although they cannot reverse any attachment loss due to periodontitis.
- Using periodontal trays to maintain dentist-prescribed medications at the source of the disease: The use of trays allows the medication to stay in place long enough to penetrate the biofilms where the micro-organism are found.

- Regular dental check-ups and professional teeth cleaning as required: Dental check-ups serve to monitor the person's oral hygiene methods and levels of attachment around teeth, identify any early signs of periodontitis, and monitor response to treatment.

Typically, dental hygienists (or dentists) use special instruments to clean (debride) teeth below the gumline and disrupt any plaque growing below the gumline. This is a standard treatment to prevent any further progress of established periodontitis. Studies show that after such a professional cleaning (periodontal debridement), microbial plaque tends to grow back to precleaning levels after about three to four months. Nonetheless, the continued stabilisation of a patient's periodontal state depends largely, if not primarily, on the patient's oral hygiene at home, as well as on the go. Without daily oral hygiene, periodontal disease will not be overcome, especially if the patient has a history of extensive periodontal disease. Periodontal disease and tooth loss are associated with an increased risk, in male patients, of cancer. Contributing causes may be high alcohol consumption or a diet low in antioxidants.

MANAGEMENT

This section from a panoramic X-ray film depicts the teeth of the lower left quadrant, exhibiting generalised severe bone loss of 30–80 per cent. The red line depicts the existing bone level, whereas the yellow line depicts where the gingiva was located originally (1–2 mm above the bone), prior to the patient developing periodontal disease. The pink arrow, on the right, points to a furcation involvement, or the loss of enough bone to reveal the location at which the individual roots of a molar begin to branch from the single root trunk; this is a sign of advanced periodontal disease. The blue arrow, in the middle, shows up to 80 per cent bone loss on tooth #21, and clinically, this tooth exhibited gross mobility. Finally, the peach oval, to the left, highlights the aggressive nature with which periodontal disease generally affects mandibular incisors. Because their roots are generally situated very close to each other, with minimal interproximal bone, and because of their location in the mouth, where plaque and calculus accumulation is greatest because of the pooling of saliva, mandibular anteriors suffer excessively. The split in the red line depicts varying densities of bone that contribute to a vague region of definitive bone height.

The cornerstone of successful periodontal treatment starts with establishing excellent oral hygiene. This includes twice-daily brushing with daily flossing. Also, the use of an interdental brush is helpful if space between the teeth allows. For smaller spaces, a product called "Soft Picks" is an excellent manual cleaning device. Persons with dexterity problems, such as arthritis, may find oral hygiene to be difficult and may require more frequent professional care and/or the use of a powered tooth brush. Persons with periodontitis must realise it is a chronic inflammatory disease and a lifelong regimen of excellent hygiene and professional maintenance care with a dentist/hygienist or periodontist is required to maintain affected teeth.

INITIAL THERAPY

Removal of microbial plaque and calculus is necessary to establish periodontal health. The first step in the treatment of periodontitis involves non-surgical cleaning below the gumline with a procedure called scaling and debridement. In the past, root planing was used (removal of cemental layer as well as calculus). This procedure involves use of specialised curettes to mechanically remove plaque and calculus from below the gumline, and may require multiple visits and local anesthesia to adequately complete. In addition to initial scaling and root planing, it may also be necessary to adjust the occlusion (bite) to prevent excessive force on teeth that have reduced bone support. Also, it may be necessary to complete any other dental needs, such as replacement of rough, plaque-retentive restorations, closure of open contacts between teeth, and any other requirements diagnosed at the initial evaluation.

REEVALUATION

Multiple clinical studies have shown non-surgical scaling and root planing are usually successful if the periodontal pockets are shallower than 4–5 mm (0.16–0.20 in). The dentist or hygienist must perform a re-evaluation four to six weeks after the initial scaling and root planing, to determine if the patient's oral hygiene has improved and inflammation has regressed. Probing should be avoided then, and an analysis by gingival index should determine the presence or absence of inflammation. The monthly reevaluation of periodontal therapy should involve periodontal charting as a better indication of the success of treatment, and to see if other courses of treatment can be identified. Pocket depths of greater than 5–6 mm (0.20–0.24 in) which remain after initial therapy, with bleeding upon probing, indicate continued active disease and will very likely lead to further bone loss over time. This is especially true in molar tooth sites where furcations (areas between the roots) have been exposed.

SURGERY

If non-surgical therapy is found to have been unsuccessful in managing signs of disease activity, periodontal surgery may be needed to stop progressive bone loss and regenerate lost bone where possible. Many surgical approaches are used in treatment of advanced periodontitis, including open flap debridement and osseous surgery, as well as guided tissue regeneration and bone grafting. The goal of periodontal surgery is access for definitive calculus removal and surgical management of bony irregularities which have resulted from the disease process to reduce pockets as much as possible. Long-term studies have shown, in moderate to advanced periodontitis, surgically treated cases often have less further breakdown over time and, when coupled with a regular post-treatment maintenance regimen, are successful in nearly halting tooth loss in nearly 85 per cent of patients.

MAINTENANCE

Once successful periodontal treatment has been completed, with or without surgery, an ongoing regimen of "periodontal maintenance" is required. This involves regular checkups and detailed cleanings every three months to prevent repopulation of periodontitis-causing microorganism, and to closely monitor affected teeth so early treatment can be rendered if disease recurs. Usually, periodontal disease exists due to poor plaque control, therefore if the brushing techniques are not modified, a periodontal recurrence is probable.

ALTERNATIVE TREATMENTS

Periodontitis has an inescapable relationship with subgingival calculus (tartar). The first step in any procedure is to eliminate calculus under the gum line, as it houses destructive anaerobic microorganisms that consume bone, gum and cementum (connective tissue) for food.

Most alternative "at-home" gum disease treatments involve injecting antimicrobial solutions, such as hydrogen peroxide, into periodontal pockets via slender applicators or oral irrigators. This process disrupts anaerobic micro-organism colonies and is effective at reducing infections and inflammation when used daily. A number of other products, functionally equivalent to hydrogen peroxide, are commercially available, but at substantially higher cost. However, such treatments do not address calculus formations, and so are short-lived, as anaerobic microbial colonies quickly regenerate in and around calculus.

Additionally, periodontitis can be treated in a non-invasive manner by means of Periostat, an FDA-approved, orally administered drug that has been shown to reduce bone loss. Its mechanism of action in part involves inhibition of matrix metalloproteinases (such as collagenase), which degrade the extracellular matrix under inflammatory conditions. This ultimately can lead to reduction of aveolar bone loss in patients with periodontal disease (as well as patients without periodontitis).

PROGNOSIS

Dentists and dental hygienists measure periodontal disease using a device called a periodontal probe. This thin "measuring stick" is gently placed into the space between the gums and the teeth, and slipped below the gumline. If the probe can slip more than 3 mm (0.12 in) below the gumline, the patient is said to have a gingival pocket if no migration of the epithelial attachment has occurred or a periodontal pocket if apical migration has occurred. This is somewhat of a misnomer, as any depth is in essence a pocket, which in turn is defined by its depth, *i.e.*, a 2-mm pocket or a 6-mm pocket. However, pockets are generally accepted as self-cleansable (at home, by the patient, with a toothbrush) if they are 3 mm or less in depth.

This is important because if a pocket is deeper than 3 mm around the tooth, at-home care will not be sufficient to cleanse the pocket, and professional care should be sought. When the pocket depths reach 6 to 7 mm (0.24 to 0.28 in) in depth, the hand instruments and cavitrons used by the dental professionals may not reach deeply enough into the pocket to clean out the microbial plaque that cause gingival inflammation. In such a situation, the bone or the gums around that tooth should be surgically altered or it will always have inflammation which will likely result in more bone loss around that tooth. An additional way to stop the inflammation would be for the patient to receive subgingival antibiotics (such as minocycline) or undergo some form of gingival surgery to access the depths of the pockets and perhaps even change the pocket depths so they become 3 mm or less in depth and can once again be properly cleaned by the patient at home with his or her toothbrush. If patients have 7-mm or deeper pockets around their teeth, then they would likely risk eventual tooth loss over the years. If this periodontal condition is not identified and the patients remain unaware of the progressive nature of the disease, then years later, they may be surprised that some teeth will gradually become loose and may need to be extracted, sometimes due to a severe infection or even pain.

According to the Sri Lankan tea labourer study, in the absence of any oral hygiene activity, approximately 10 per cent will suffer from severe periodontal disease with rapid loss of attachment (>2 mm/year). About 80 per cent will suffer from moderate loss (1–2 mm/year) and the remaining 10 per cent will not suffer any loss.

EPIDEMIOLOGY

Periodontitis is very common, and is widely regarded as the second most common disease worldwide, after dental decay, and in the United States has a prevalence of 30–50 per cent of the population, but only about 10 per cent have severe forms. Chronic periodontitis effects about 750 million people or about 10.8 per cent of the population as of 2010. Like other conditions intimately related to access to hygiene and basic medical monitoring and care, periodontitis tends to be more common in economically disadvantaged populations or regions. Its occurrence decreases with higher standard of living. In Israeli population, individuals of Yemenite, North-African, South Asian, or Mediterranean origin have higher prevalence of periodontal disease than individuals from European descent.

IN OTHER ANIMALS

Periodontal disease is the most common disease found in dogs and affects more than 80 per cent of dogs aged three years or older. Its prevalence in dogs increases with age, but decreases with increasing body weight; *i.e.*, toy and miniature breeds are more severely affected. Systemic disease may develop because the gums are very vascular (have a good blood supply). The blood stream carries these anaerobic micro-organisms, and they are filtered out by the kidneys and liver, where they may colonise and create microabscesses. The micro-organisms traveling through the blood may also attach to the heart valves, causing vegetative endocarditis (infected heart valves). Additional diseases that may result from periodontitis include chronic bronchitis and pulmonary fibrosis.

STAPHYLOCOCCUS

Staphylococcus (from the Greek: σταφνλη, *staphylç*, "grape" and κοκκοζ, *kókkos*, "granule") is a genus of Gram-positive bacteria. Under the microscope, they appear round (cocci), and form in grape-like clusters. The *Staphylococcus* genus includes at least 40 species. Of these, nine have two subspecies and one has three subspecies. Most are harmless and reside normally on the skin and mucous membranes of humans and other organisms. Found worldwide, they are a small component of soil microbial flora.

TAXONOMY

The taxonomy is based on 16s rRNA sequences, and most of the staphylococcal species fall into 11 clusters:

1. S. aureus group – S. aureus, S. simiae
2. S. auricularis group – S. auricularis
3. S. carnosus group – S. carnosus, S. condimenti, S. massiliensis, S. piscifermentans, S. simulans
4. S. epidermidis group – S. capitis, S. caprae, S. epidermidis, S. saccharolyticus
5. S. haemolyticus group – S. devriesei, S. haemolyticus, S. hominis
6. S. hyicus-intermedius group – S. chromogenes, S. felis, S. delphini, S. hyicus, S. intermedius, S. lutrae, S. microti, S. muscae, S. pseudintermedius, S. rostri, S. schleiferi
7. S. lugdunensis group – S. lugdunensis
8. S. saprophyticus group – S. arlettae, S. cohnii, S. equorum, S. gallinarum, S. kloosii, S. leei, S. nepalensis, S. saprophyticus, S. succinus, S. xylosus
9. S. sciuri group – S. fleurettii, S. lentus, S. sciuri, S. stepanovicii, S. vitulinus
10. S. simulans group – S. simulans
11. S. warneri group – S. pasteuri, S. warneri

A twelfth group – that of *S. caseolyticus* – has now been moved to a new genus *Macrococcus*, the species of which are currently the closest known relatives of the *Staphylococci*.

SUBSPECIES

- *S. aureus* subsp. *aureus*
- *S. aureus* subsp. *anaerobius*
- *S. capitis* subsp. *capitis*
- *S. capitis* subsp. *urealyticus*
- *S. carnosus* subsp. *carnosus*
- *S. carnosus* subsp. *utilis*
- *S. cohnii* subsp. *cohnii*
- *S. cohnii* subsp. *urealyticus*
- *S. equorum* subsp. *equorum*
- *S. equorum* subsp. *linens*
- *S. hominis* subsp. *hominis*
- *S. hominis* subsp. *novobiosepticus*
- *S. saprophyticus* subsp. *bovis*
- *S. saprophyticus* subsp. *saprophyticus*
- *S. schleiferi* subsp. *coagulans*
- *S. schleiferi* subsp. *schleiferi*
- *S. sciuri* subsp. *carnaticus*
- *S. sciuri* subsp. *rodentium*
- *S. sciuri* subsp. *sciuri*
- *S. succinus* subsp. *casei*
- *S. succinus* subsp. *succinus*

NOTES

As with all generic names in binomial nomenclature, *Staphylococcus* is capitalised when used alone or with a specific species. Also, the abbreviations *Staph* and *S.* when used with a species (*S. aureus*) are correctly italicised and capitalised (though often errors in this are seen in popular literature). However, *Staphylococcus* is not capitalised or italicised when used in adjectival forms, as in a staphylococcal infection, or as the plural (staphylococci).

The *S. saprophyticus* and *S. sciuri* groups are generally novobiocin-resistant, as is *S. hominis* subsp. *novobiosepticus*.

Members of the *S. sciuri* group are oxidase-positive due to their possession of the enzyme cytochrome c oxidase. This group is the only clade within the *Staphylococci* to possess this gene.

The *S. sciuri* group appears to be the closest relations to the genus *Macrococcus*.

Staphylococcus pulvereri has been shown to be a junior synonym of *Staphylococcus vitulinus*.

Within these clades, the *S. haemolyticus* and *S. simulans* groups appear to be related, as do the *S. aureus* and *S. epidermidis* groups.

S. lugdunensis appears to be related to the *S. haemolyticus* group.

S. croceolyticus may be related to *S. haemolyticus*, but this needs to be confirmed.

The taxonomic position of *S. croceolyticus*, *S. leei*, *S. lyticans* and *S. pseudolugdunensis* has yet to be clarified. The published descriptions of these species do not appear to have been validly published to date (2010).

BIOCHEMICAL IDENTIFICATION

Assignment of a strain to the genus *Staphylococcus* requires it to be a Gram-positive coccus that forms clusters, produces catalase, has an appropriate cell wall structure (including peptidoglycan type and teichoic acid presence) and G + C content of DNA in a range of 30–40 mol per cent. *Staphylococcus* species can be differentiated from other aerobic and facultative anaerobic, Gram-positive cocci by several simple tests. *Staphylococcus* spp. are facultative anaerobes (capable of growth both aerobically and anaerobically). All species grow in the presence of bile salts. It was believed that all species were coagulase-positive however it is now known that not all *Staphylococcus* are coagulase positive. Growth can also occur in a 6.5 per cent NaCl solution. On Baird Parker medium, *Staphylococcus* spp. grow fermentatively, except for *S. saprophyticus*, which grows oxidatively. *Staphylococcus* spp. are resistant to bacitracin (0.04 U disc: resistance = <10mm zone of inhibition) and susceptible to furazolidone (100μg disc: resistance = <15mm zone of inhibition). Further biochemical testing is needed to identify to the species level. When the bacterium divides it divides along two axes, so forming clumps of bacteria. This is as opposed to streptococci which divide along one axis and so form chains (strep. meaning twisted or pliant).

COAGULASE PRODUCTION

One of the most important phenotypical features used in the classification of staphylococci is their ability to produce coagulase, an enzyme that causes blood clot formation.

Six species are currently recognised as being coagulase-positive:

1. *S. aureus*,
2. *S. delphini*,
3. *S. hyicus*,
4. *S. intermedius*,
5. *S. lutrae*,*S. pseudintermedius* and
6. *S. schleiferi* subsp. *coagulans*.

These species belong to two separate groups – the *S. aureus* (*S. aureus* alone) group and the *S. hyicus-intermedius* group (the remaining five). *S. aureus* can also be found as being coagulase-negative.

A seventh species has also been described – *Staphylococcus leei* – from patients with gastritis. *S. aureus* is coagulase-positive, meaning it produces coagulase. However, while the majority of *S. aureus* strains are coagulase-positive, some may be atypical in that they do not produce coagulase. *S. aureus* is catalase-positive (meaning that it can produce the enzyme catalase) and able to convert hydrogen peroxide (H_2O_2) to water and oxygen, which makes the catalase test useful to distinguish staphylococci from enterococci and streptococci.

S. pseudintermedius inhabits and sometimes infects the skin of domestic dogs and cats. This organism, too, can carry the genetic material that imparts multiple bacterial resistance. It is rarely implicated in infections in humans, as a zoonosis.

S. epidermidis, a coagulase-negative species, is a commensal of the skin, but can cause severe infections in immune-suppressed patients and those with central venous catheters. *S. saprophyticus*, another coagulase-negative species that is part of the normal vaginal flora, is predominantly implicated in genitourinary tract infections in sexually active young women. In recent years, several other *Staphylococcus* species have been implicated in human infections, notably *S. lugdunensis*, *S. schleiferi*, and *S. caprae*. Common abbreviations for coagulase-negative staphylococcus species are CoNS and CNS.

GENOMICS AND MOLECULAR BIOLOGY

The first *S. aureus* genomes to be sequenced were those of N315 and Mu50 in 2001. Many more complete *S. aureus* genomes have been submitted to the public databases, making it one of the most extensively sequenced bacteria. The use of genomic data is now widespread and provides a valuable resource for researchers working with *S. aureus*. Whole genome technologies, such as sequencing projects and microarrays, have shown an enormous variety of *S. aureus* strains. Each contains different combinations of surface proteins and different toxins. Relating this information to pathogenic behaviour is one of the major areas of staphylococcal research. The development of molecular typing methods has enabled the tracking of different strains of *S. aureus*. This may lead to better control of outbreak strains. A greater understanding of how the staphylococci evolve, especially due to the acquisition of mobile genetic elements encoding resistance and virulence genes is helping to identify new outbreak strains and may even prevent their emergence.

The widespread incidence of antibiotic resistance across various strains of *S. aureus*, or across different species of *Staphylococcus* has been attributed to horizontal gene transfer of genes encoding antibiotic/metal resistance and virulence. A recent study demonstrated the extent of horizontal gene transfer among *Staphylococcus* to be much greater than previously expected, and encompasses genes with functions beyond antibiotic resistance and virulence, and beyond genes residing within the mobile genetic elements. Various strains of *Staphylococcus* are available from biological research centres, such as the National Collection of Type Cultures (NCTC).

HOST RANGE

Members of the genus *Staphylococcus* frequently colonise the skin and upper respiratory tracts of mammals and birds. Some species specificity has been observed in host range, such that the *Staphylococcus* species observed on some animals appear more rarely on more distantly related host species.

Some of the observed host specificity includes:

- *S. arlattae* – chickens, goats
- *S. auricularis* – deer, dogs, humans
- *S. capitis* – human
- *S. caprae* – goats, humans
- *S. cohnii* – chickens, humans
- *S. delphini* – dolphins
- *S. devriesei* – cattle
- *S. epidermiditis* – humans
- *S. equorum* – horses
- *S. felis* – cats
- *S. fleurettii* – goats
- *S. gallinarum* – chickens, goats, pheasants
- *S. haemolyticus* – humans, *Cercocebus*, *Erythrocebus*, *Lemur*, *Macca*, *Microcebus*, *Pan*
- *S. hyicus* – pigs
- *S. leei* – humans
- *S. lentus* – goats, rabbits, sheep
- *S. lugdunensis* – humans, goats
- *S. lutrae* – otters
- *S. microti* – voles (*Microtus arvalis*)
- *S. nepalensis* – goats
- *S. pasteuri* – humans, goats
- *S. pettenkoferi* – humans
- *S. pseudintermedius* – dogs
- *S. rostri* – pigs
- *S. schleiferi* – humans
- *S. sciuri* – humans, dogs, goats
- *S. simiae* – South American squirrel monkeys (*Saimiri sciureus*)
- *S. simulans* – humans
- *S. warneri* – humans, Cercopithecoidea, Pongidae
- *S. xylosus* – humans

STAPHYLOCOCCAL INFECTION OF BACTERIA

Staphylococcus is a Gram-positive bacteria which includes several species that can cause a wide variety of infections in humans and other animals through infection or the production of toxins. Staphylococcal toxins are a common cause of food poisoning, as they can be produced in improperly-stored food. Staphylococci are also known to be a cause of bacterial pink eye.

COAGULASE-POSITIVE

The main coagulase-positive staphylococcus is *Staphylococcus aureus*, although not all strains of *Staphylococcus aureus* are coagulase positive.

These bacteria can survive on dry surfaces, increasing the chance of transmission. *S. aureus* is also implicated in toxic shock syndrome; during the 1980s some tampons allowed the rapid growth of *S. aureus*, which released toxins that were absorbed into the bloodstream. Any *S. aureus* infection can cause the staphylococcal scalded skin syndrome, a cutaneous reaction to exotoxin absorbed into the bloodstream. It can also cause a type of septicaemia called pyaemia. The infection can be life-threatening. Problematically, Methicillin-resistant *Staphylococcus aureus* (MRSA) has become a major cause of hospital-acquired infections, and is being recognised with increasing frequency in community-acquired infections.

ETYMOLOGY

The generic name *Staphylococcus* is derived from the Greek word "staphyle" meaning a bunch of grapes, and "kokkos" means granule. The bacteria, when seen under a microscope appear like a branch of grapes or berries.

Main Staphylococcus aureus infections

Type	Examples
Localised skin infections	• Stye and other small, superficial abscesses in sweat or sebaceous glands • Subcutaneous abscesses (boils) around foreign bodies • Large, deep, infections (carbuncles) possibly causing bacteremia
Diffuse skin infection	• Impetigo
Deep, localised infections	• Acute and chronic osteomyelitis • Septic arthritis
Other infections	• Acute infective endocarditis • Septicemia • Necrotising pneumonia
Toxinoses	• Toxic shock syndrome • Gastroenteritis • Scalded skin syndrome

COAGULASE-NEGATIVE

- *S. epidermidis*, a coagulase-negative staphylococcus species, is a commensal of the skin, but can cause severe infections in immune-suppressed patients and those with central venous catheters.

- *S. saprophyticus*, another coagulase-negative species that is part of the normal vaginal flora, is predominantly implicated in genitourinary tract infections in sexually-active young women.
- In recent years, several other staphylococcal species have been implicated in human infections, notably *S. lugdunensis*, *S. schleiferi*, and *S. caprae*.

STREPTOCOCCUSOF BACTERIA

Streptococcus is a genus of spherical Gram-positive bacteria belonging to the phylum Firmicutes and the lactic acid bacteria group. Cellular division occurs along a single axis in these bacteria, and thus they grow in chains or pairs, hence the name — from Greek στρεπτοζ *streptos*, meaning easily bent or twisted, like a chain (twisted chain). Contrast this with staphylococci, which divide along multiple axes and generate grape-like clusters of cells. Most streptococci are oxidase- and catalase-negative, and many are facultative anaerobes.

In 1984, many organisms formerly considered *Streptococcus* were separated out into the genera *Enterococcus* and *Lactococcus*. There are currently over 50 species recognised in this genus.

In addition to streptococcal pharyngitis (strep throat), certain *Streptococcus* species are responsible for many cases of pink eye, meningitis, bacterial pneumonia, endocarditis, erysipelas and necrotising fasciitis (the 'flesh-eating' bacterial infections).

However, many streptococcal species are non-pathogenic, and form part of the commensal human microbiome of the mouth, skin, intestine, and upper respiratory tract. Furthermore, streptococci are a necessary ingredient in producing Emmentaler ("Swiss") cheese.

Species of *Streptococcus* are classified based on their hemolytic properties. Alpha hemolytic species cause oxidisation of iron in hemoglobin molecules within red blood cells, giving it a greenish colour on blood agar.

Beta hemolytic species cause complete rupture of red blood cells. On blood agar, this appears as wide areas clear of blood cells surrounding bacterial colonies. Gamma-hemolytic species cause no hemolysis.

Beta-hemolytic streptococci are further characterised via Lancefield serotyping, which describes specific carbohydrates present on the bacterial cell wall.

There are 20 described serotypes, named Lancefield groups A to V (excluding I and J). In the medical setting, the most important groups are the alpha-hemolytic streptococci *S. pneumoniae* and *Streptococcus* Viridans-group, and the beta-hemolytic streptococci of Lancefield groups A and B (also known as "Group A strep" and "Group B strep").

ALPHA-HEMOLYTIC

Pneumococci

- *S. pneumoniae* (sometimes called *Pneumococcus*), is a leading cause of bacterial pneumonia and occasional etiology of otitis media, sinusitis, meningitis and peritonitis.

Inflammation is thought to be the major cause of how pneumococcus causes disease, hence the tendency of diagnoses associated with it to involve inflammation.

The Viridans Group: Alpha-hemolytic

- Streptococcus viridans is a pseudotaxonomic non-Linnean term for a large group of commensal streptococcal bacteria that are either a-hemolytic, producing a green colouration on blood agar plates (hence the name "viridans", from Latin "", green), or non-hemolytic. They possess no Lancefield antigens.

Group A

S. pyogenes, also known as Group A *Streptococcus* (GAS), is the causative agent in a wide range of Group A streptococcal infections. These infections may be non-invasive or invasive. The non-invasive infections tend to be more common and less severe. The most common of these infections include *streptococcal pharyngitis* (strep throat) and impetigo. Scarlet fever is also a non-invasive infection, but has not been as common in recent years.

The invasive infections cause by Group A â-hemolytic streptococcus tend to be more severe and less common. This occurs when the bacterium is able to infect areas where it is not usually found, such as the blood and the organs. The diseases that may be caused as a result of this include streptococcal toxic shock syndrome (STSS), necrotising fasciitis (NF), pneumonia, and bacteremia. Additional complications may be caused by GAS, namely acute rheumatic fever and acute glomerulonephritis. Rheumatic fever, a disease that affects the joints, kidneys, and heart valves, is a consequence of untreated strep A infection caused not by the bacterium itself. Rheumatic fever is caused by the antibodies created by the immune system to fight off the infection cross-reacting with other proteins in the body. This "cross-reaction" causes the body to essentially attack itself and leads to the damage above. Globally, GAS has been estimated to cause more than 500,000 deaths every year, making it one of the world's leading pathogens. Group A *Streptococcus* infection is generally diagnosed with a Rapid Strep Test or by culture.

Group B

S. agalactiae, or GBS, causes pneumonia and meningitis in neo-nates and the elderly, with occasional systemic bacteremia. They can also colonise the intestines and the female reproductive tract, increasing the risk for premature rupture of membranes during pregnancy, and transmission of the organism to the infant. The American College of Obstetricians and Gynecologists, American Academy of Pediatrics and the Centers for Disease Control recommend that all pregnant women between 35 and 37 weeks gestation should be tested for GBS. Women who test positive should be given prophylactic antibiotics during labour, which will usually prevent transmission to the infant. In the UK, clinicians have been slow to implement the same standards as the US, Australia and Canada. In the UK, only 1 per cent of maternity units test for the presence of Group B *Streptococcus*. Although the Royal College of Obstetricians and Gynaecologists issued risk-based guidelines in 2003 (due for review 2006), the implementation of these guidelines has been patchy. As a result, over 75 infants in the UK die each year of GBS-related disease, and another 600 or so suffer serious infection, most of which could be prevented; however, this is yet to be substantiated by randomised, controlled trial in the UK setting and, given the evidence for the efficacy of testing and treating from other countries, it may be that the large-scale trial necessary would receive neither funding nor ethics approval.

Group C

This group includes *S. equi*, which causes strangles in horses, and *S. zooepidemicus* - *S. equi* is a clonal descendent or biovar of the ancestral *S. zooepidemicus* - which causes infections in several species of mammals, including cattle and horses. *Streptococcus dysgalactiae* is also a member of Group C, β-haemolytic streptococci that can cause pharyngitis and other pyogenic infections similar to Group A streptococci.

Group D (Enterococci)

Many former Group D streptococci have been reclassified and placed in the genus *Enterococcus* (including *Enterococcus faecalis*, *Enterococcus faecium*, *Enterococcus durans*, and *Enterococcus avium*). For example, *Streptococcus faecalis* is now *Enterococcus faecalis*.

The remaining non-enterococcal Group D strains include *Streptococcus bovis* and *Streptococcus equinus*. Non-hemolytic streptococci rarely cause illness. However, weakly hemolytic group D beta-hemolytic streptococci and *Listeria monocytogenes* (which is actually a Gram-positive bacillus) should not be confused with non-hemolytic streptococci.

Group F Streptococci

Group F streptococci were first described in 1934 by Long and Bliss amongst the "minute haemolytic streptococci". They are also known as

Streptococcus anginosus (according to the Lancefield classification system) or as members of the *S. milleri* group (according to the European system).

Group G Streptococci

These streptococci are usually, but not exclusively, beta-hemolytic. *Streptococcus canis* is an example of a GGS which is typically found on animals, but can cause infection in humans.

Group H Streptococci

Group H streptococci is a very rare infection caused by the herptitis bacteria found in medium sized canines. Group H streptococci rarely causes illness unless a human has direct contact with the mouth of a canine. One of the most common ways this can be spread is human to canine, mouth-to-mouth contact. However, canine may lick the humans hand and infection can be spread as well.

MOLECULAR TAXONOMY AND PHYLOGENETICS

Fig. Phylogenetic Tree of *Streptococcus* Species, based on Data from PATRIC. 16S Groups are Indicated by Brackets and their Key Members.

Streptococci have been divided into six groups on the basis of their 16S rDNA sequences:

1. Anginosus,
2. Bovis,
3. Mitis,
4. Mutans,
5. *Pyogenes* and
6. *Salivarius*.

The 16S groups have been confirmed by whole genome sequencing. The important pathogens *S. pneumoniae* and *S. pyogenes* belong to the Mitis and Pyogenes groups, respectively, while the causing agent of dental caries,*Streptococcus mutans*, is basal to the *Streptococcus* group.

GENOMICS

The genomes of hundreds of species have been sequenced. Most Streptococcus genomes are 1.8 tp 2.3 Mb in size and encode 1,700 to 2,300 proteins. Some important genomes are listed in the table. The four species shown in the table (*pyogenes, agalactiae, pneumoniae*, and *mutans*) have an average pairwise protein sequence identity of about 70 per cent.

STREPTOCOCCAL INFECTION IN POULTRY

Streptococcus species are the cause of opportunistic infections in poultry leading to acute and chronic conditions in affected birds. Disease varies according to the Streptococcal species but common presentations include septicaemia, peritonitis, salpingitis and endocarditis.

Common species affecting poultry include:

- *S. gallinaceus* in broiler chickens
- *S. gallolyticus* which is a pathogen of racing pigeons and turkey poults
- *S. dysgalactiae* in broiler chickens
- *S. mutans* in geese
- *S. pluranimalium* in broiler chickens
- *S. equi subsp. zooepidemicus* in chickens and turkeys
- *S. suis* in psittacine birds

DIAGNOSIS

Post-mortem findings include friable internal organs, abdominal effusion and evidence of sepsis in the joints, heart valves and brain. Bacteria can usually be cultured from tissues collected at necropsy or identified by microscope examination.

TREATMENT AND CONTROL

The organism should be cultured and antibiotic sensitivity should be determined before treatment is started. Amoxycillin is usually effective in treating streptococcal infections. Biosecurity protocols and good hygiene are important in preventing the disease. Vaccination is available against *S. gallolyticus* and can also protect pigeons.

4

Gene Therapy

INTRODUCTION

Gene therapy is a novel approach to treat, cure or ultimately prevent disease by changing the expression of a person's gene. Or The treatment of any disease by the introduction of genes into affected cells to provide gene functions that are deficient because of genetic or acquired deficiency is known as gene therapy. Gene therapy is in its infancy, and current gene therapy is primarily experimental, with most human clinical trials only in research stages. The development of a therapeutic agent for humans usually passes through four clearly defined levels of study and testing before it can be sanctioned for use in regular practice.

This makes sure that any therapeutic agents should not show any harmful effects or side effects on the host or receiving organisms.

- *Preclinical trials*: Animal testing of the therapeutic agent to determine a rationale for therapeutic use, toxicity's, dose details, and risks involved.
- *Phase I*: A small human trial generally involving dose escalation and focusing on safety and pharmacology of therapeutic agent.
- *Phase II*: An expanded human trial with the focus on efficacy in a particular patient population.
- *Phase III*: A large human trial intended to establish a definitive role for a drug or biologic.

The same things should also be followed by the gene therapy technology. Even though, there is no new or synthetic organics present in the gene therapy, the cells used for the process may creates unexpected problems.

Current practices of symptomatic treatment for genetic disorder are:

- Replacement therapy for missing factor. Eg: coagulation factor for hemophilia.
- Long term blood transfusion for thalassaemia.
- Replacement with immunoglobulins for children with congenital hypogammaglobulinemia.

- Hormone therapy for certain diseases. Eg: Growth hormone for dwarfism, insulin injection for diabetics.
- Replacement of missing enzymes. Eg: Gauchers disease, ADA deficiency.
- Diet control prevents accumulation of toxic metabolites. Eg: Phenylketonuria, galactosemia etc.
- Bone marrow transplantation to correct blood disorders. Eg: SCID viii. Cofactor responsive metabolic disorders. Eg: methylmalonic acidaemia, homocystinuria colatamine, multiple acyl-coA dehydrogen, deficiency-riboflavin etc.

Symptomatic disease treatments show very low rate of success. It is completely successful in eight diseases (12 per cent of total), moderately successful in 14 per cent of the total and not cured in rest. Majority of the genetic diseases are not treated by regular conventional therapeutic agents such as synthetic organic drugs or naturally isolated supplements. This makes the treatment processes as only a short term solution or cure.

This treatment can also cause a large number of side effects in an accumulative manner. And this will not be able to provide the complete curing effect and makes the patient to be under continuous medication or life long treatment. In such long term treatments regular changes of therapeutic agents should be made as a regular practice. This has a potential to alter the entire physiology of the host and thus leads to unrepairable damage to the various organs of the body.

WHAT IS GENE THERAPY?

Every cell of the human body contains 2 metres of DNA. DNA itself is made of the four bases adenine, guanine, cytosine and thymine. Three bases in line are called a codon and encode one of the 20 amino acids used in the human body. Lots of amino acids together form a protein which has a particular function in the body. The DNA is packed into chromosomes. A human being has 46 chromosomes, 23 from the mother 23 from the father. They contain the same genes more or less at the same place so that in every ~~no~~ germ line cell are two copies of a particular gene. Germ line cells we call cells like sperms or eggs which only contain 23 chromosomes, coming together they form a new cell with 46 chromosomes witch has the capacity to develop into a new human being.

Inherited diseases such as cystic fibrosis are caused by a non-functional protein which is caused by a mutation in the DNA which leads for example to a loss of one or more amino acids or to a wrong amino acid in the protein chain. There are lots of different types of mutation like frame shift mutations, deletions, non-sense mutation and so on and so forth. In the case of cystic fibrosis the mutation is a deletion, which means one ore more DNA bases are lost and the DNA is shorter than normal.

There are different kinds of inherited diseases. Autosomal recessive diseases are located on one of the 22 autosomal chromosomes (not on the x or y chromosome) and there must be two wrong copies of the gene to cause the disease. If there is one copy of the wrong gene the individual is heterozygote and has a normal phenotype but can pass the wrong gene to his offspring. Gene therapy tries to replace or support mutated genes by wild-type (normal) ones.

There are different ways to do this. Germ-line therapy would try to put a wild-type gene directly into a germ-line or embryonic cell so that all cells which develop from this cell carry the wild type gene. Somatic gene therapy tries to correct a mutated gene by introducing healthy genes in the affected cells either by removing cells "correcting" their DNA and reintroducing them, or by using viruses, bacteria or liposome as vectors to introduce wild-type DNA into an affected cell. In that case the DNA of a virus for example is treated so the virus is not dangerous anymore and contains the wild-type gene of interest. But the virus still can attack cells and introduce its own DNA in the cell; by doing so it introduces the wild-type gene in the host cell which should lead to a normal gene product (protein). One of the problems with this method is that the body often reacts with an immune deficiency to the virus.

Another way to introduce DNA in cells is to use liposome, small drops of fat, which you can put the DNA in. The liposomes are taken up through endocytosis by the cell. There is no immune deficiency but bringing the DNA from the cytosol into the nucleus remains a problem.

VIRAL VECTORS

Adenovirus vectors

Adenoviruses are non-enveloped double-stranded DNA viruses. Their genome contains 30-40 kb. Adenoviruses are able to transduce a variety of cell types, including non-dividing cells. Because it is relatively easy to genetically manipulate their genome, adenoviruses were the first vectors used in clinical trials. There are first and second generation adenoviruses. Viruses which encode early transcripts, for example proteins responsible for transcriptional activation, are called the first generation. Viruses encoding later transcripts belong to the second generation. The virus genome has different regions called E1A (early region, encodes two proteins for transcriptional activation), E1B (prevents apoptosis induced by E1A), E2A (encodes the DNA binding protein), E2B (encodes DNA polymerase and pre-terminal Protein), E3 (encodes for proteins involved in evasion of the host cell) and E4 (encodes for proteins which inhibit cellular mRNA export from the nucleus). Vectors deleted in the E2B region show drastically reduced viral replication and thus a reduction in immune response. Adenovirus vector with a deletion in the E4 region appear to be particularly susceptible to down-regulation of trans-gene expression. To prevent virus replication and as a consequent host cell lysis, most of the E1 region is deleted when, generating adenovirus vectors in the laboratory.

"The first human gene therapy trial for cystic fibrosis was commenced in New York April 1993. Four patients had 2×10^6 -2×10^9 plaque-forming units of the Ad2 (adenovirus 2) CFTR encoding vector administered in liquid form, initially to the nasal epithelium, the one day later to a lower lobe bronchus via a bronchoscope." (1) CFTR protein expression increased for 10 days after administration but unfortunately, dose-limiting systemic and pulmonary inflammation occurred.

A similar experiment was made soon afterwards from a group at the North Carolina University. They used a HAdV-2(human adenovirus scrotype 2) vector with deletions of the E1A, E3 and partial E1b region. 14 individuals with cystic fibrosis were treated using 3-monthly dosing cycles. There was less local inflammation, but might be only because the dose was not as high as in the first experiment. However vector-derived CFTR expression persisted only 4-30 days following the first dose. After the second does expression was reduced, and following the third dose expression was undetectable.

"This persistent poor efficacy of adenovirus vectors to deliver the CFTR gene to terminally differentiated respiratory epithelium may be partly due to the CAR(Coxsackie B and adenovirus receptor) being predominantly located on the basolatheral rather than the luminal surface of human respiratory cells. Non-clinical studies have thus explored the potential of pseudo typing the adenovirus vectors to target other more suitable cell surface receptors." (1)

However clinical research into this vector-system has been reduced after the death of Jesse Gelsinger who died because of an overwhelming immune response to an adenovirus vector.

Adeno-associated virus (AAV) vectors

AAV is a parvovirus that needs a helper virus to replicate. Usually this helper virus is an adenovirus. AAV does not cause any illness in humans. If the helper virus is absent, the AAV can produce a latent infection by stable integration of its genome. It integrates into the human chromosome 19. A recombinant AAV without the replicase gene, does not integrate at this specific side. The first phase in a clinical trial, where 10 cystic fibrosis patients were treated with an AAV containing the CFTR gene sequence, showed no evidence of inflammation. Gene transfer was achieved for 1 in 10 respiratory epithelia cells for up to 10 weeks but no vector-derived RNA could be detected in any sample. "This poor expression was attributed to lack of an efficient due promoter due to AAV only having a packing capacity of approx. 4.8 kb. A further concern was the rise in scrum-neutralizing antibodies to AAV observed in the study."(1)

These results suggested that gene transfer and "expression" will decrease with any repetition of the gene therapy in a patient. A clinical trial with 37 cystic fibrosis patients was made. They were treated at 4 weeks intervals with either 10^{13} tg AAV- CFTR particles or placebo.

For the first time in any trial of gene therapy in cystic fibrosis, there was a significant difference between the placebo and the tgAAV- CFTR group. The treated group showed better results in respiratory functions and reduced levels of pro-inflammatory cytokine interleukin-8. However the effectiveness was reduced with every repetition of the treatment, associated with a rise in neutralizing antibodies. There is further work done to improve the effectiveness of this promising vector.

Other viral vectors

The most promising other virus vector systems are the lentiviral vectors and parainfluenza virus vectors type 3. Lentiviral vectors integrate their genome into the host DNA. This offers the possibility of providing sustained expression of the therapeutic gene. However the problem remains the same as with adenovirus and AAV as well. They are hampered by a lack of suitable receptors on the apical of bronchial epithelia cells.

Another problem caused by lentiviral or retroviral vectors is that they integrate randomly into the genome. Two children with severe combined immunodeficiency were treated with retrovirus-mediated gene therapy and developed a T-cell leukaemia possibly as a result of gene inactivation. Such problems have to be solved prior to a clinical trial in cystic fibrosis patients.

The advantage of parainfluenza virus type 3 is that it targets ciliated airway epithelial cells from the apical surface. As a vector containing the CFTR gene it efficiently corrects abnormalities seen in cystic fibrosis bronchial epithelial cells in tissue culture. Research is going on to find forms that can safely deliver genes to the human airway epithelia cells *in vivo.*

NON- VIRAL GENE DELIVERY

Cationic Liposomes

A liposome is a lipid bilayer forming a kind of a balloon that coats foraging DNA, containing for example the CFTR gene.

Cationic liposomes have positively charged side chains. This side chains interact with the negatively charged DNA. The hydrophobic lipid portion of the liposome enhances fusion with the host cell membrane. One advantage to viral vectors is that there is no limitation to the length of DNA which is incorporated. Lipofectin was the first commercially available cationic liposome. It consists of DOTMA (N-(1-(2,3-dioletyloxy)propyl)-N,N,N-trimethyla-mmonium chloride) and the colipid DOPE (diolcoyl phosphatidyle-thanolamine), is used commercially as Lipofectamin, which contains a spermine group. Due to the risk of toxicity related with this group Lipofectamin is unsuitable for use *in vivo*. There have been lots of studies *in vitro*. The most relevant liposome used in cystic fibrosis clinical trials today is EDMPC (1,2-dimiristoyl-P-O-ethylphosphatidilcholine) mixed with cholesterol. In one of these, 16 cystic fibrosis patients were treated either with a cationic liposome containing a CFTR plasmid DNA or the lipid alone. Unfortunately seven of the eight patients in the treated group developed influenza-like symptoms. These symptoms were not seen with lipid alone. The authors suggested that the high unmethylated cytosine and guanine content of bacterial DNA may have been responsible. But there were encouragingly effects as well.

Treated individuals showed chloride conductance significantly restored to 25 per cent of normal values, decreased lung inflammatory markers and decreased *Ps.aeruginosa* adherence. Other promising trials were made but although there has been no evidence of a reduction in effectiveness in repeat dosing or of a severe immune response, up to now levels of gene transfer are to low to result in clinical benefit. Ongoing work is based on a detailed understanding of the barriers facing plasmid DNA delivery by cationic liposomes. One way is the entry via the endocytotic pathway. But there remains the problem of transporting the plasmid DNA through the nuclear pore into the nucleus of a non-dividing cell. Small molecules of up to 10 nm can pass through by passive diffusion. Larger molecules require NLS (a peptide) for active transport through the pore.

DISADVANTAGES OF THE ABOVE THERAPY PROCESSES

- High cost is involved in the method of practice.
- Need of continuous treatment throughout the life.
- Danger of transmission of AIDS virus and other pathogenic agents during blood transfusion.
- High iron contents after repeated blood transfusions in transgenic patients has side effects. Hence, at present lot of emphasis is given to find out a better and permanent solution for the genetic disorders that are found to be effective in all respect. For this purpose the gene therapy is found out to be the most effective therapeutic method.

THE ETHICS OF GENE THERAPY

Gene therapy has become an increasingly pertinent subject in recent times. The technological advancement in the field of genetics has raised a number of philosophical, religious, and ethical questions in the process making scientific strides. A great number of such questions remain unanswered. The alteration of genetic material has the ability to resolve a great deal of medical problems if it is correctly applied. If misapplied, however, such technology could conceivably cause further social and ethical problems for the human race at large. Before launching into an ethical and religious discussion of gene therapy, it is important to understand precisely the techniques and approaches gene therapy employs. The most common methods currently available for genetic alteration make use of retroviruses. In this process, a specific set of three genes is knocked out of the genome of the natural retrovirus (Cummings & Klug, 2003). Next, the cloned and amplified human gene is inserted in the place of the knockout virus gene. The genes removed from the retrovirus eliminate its fecundity. Essentially, the virus is made impotent and not allowed to reproduce as it would naturally. Then, the affected human tissue is infected with the genetically altered retrovirus. The transfer of genetic material into the nucleus of human cells is facilitated by chemical assistance. Effectively, the recombinant virus DNA is injected into the nucleus of the human cell and integrated into the genome. Once the novel gene has established a presence in the human genome, it is transcribed and translated as if it were a gene that had always been a part of the genetic composition of that individual. The protein products of the genetic alteration are then able to displace the products of disease causing genes. In this way, the diseases and disorders that are caused by genetic influences can be ameliorated through gene therapy (Cummings & Klug, 2003). The first successful gene therapy procedure was conducted in 1990 on a child with severe combined immunodeficiency (SCID) (Cummings & Klug, 2003).This disorder affects the immune system, rendering it almost completely non-functional. For this reason, children with SCID rarely survive to puberty because they are seriously affected by what would normally be minor infections. The gene that causes SCID was isolated, and it was determined that the disorder was monogenic (having a mutation at a single gene). The responsible gene coded for a protein that was absolutely necessary for proper immune system functioning. Through the use of a retrovirus, the child made a reasonable recovery and leads a normal life today (Cummings & Klug, 2003). However, gene therapy is by no means a perfect science; many trials were not successful in even partially treating patients. Not every procedure was met with perfect success because the human cells are not always infected by the retrovirus, and transduction of genetic information does not always occur (Boylan & Brown, 2001). However, the methods and procedures by which genetic transfer occur are improving constantly by way of intense scientific research.

There are two ways in which gene therapy can be administered. Firstly, it is possible to conduct therapy on the living body tissue. Such methods are commonly referred to as somatic gene therapy (Annas & Elias, 1992). By way of this technique, the genetic changes that occur via the retrovirus vector are specific to the individual undertaking the procedure. Furthermore, the alteration of genetic material is not transmitted to future generations. Conversely, the alternative approach to somatic gene therapy is germ-line gene therapy. Germ-line gene therapy affects the gametes of the individual undergoing the procedure (Annas & Elias, 1992). As such, the future progeny of that individual will consequently bear the genetic changes made during the procedure. Germ-line gene therapy is by far the more permanent and contentious of the two types of therapy. The permanent "improvement" of the genetic composition of the human race is obviously a far more controversial subject because its application has a greater affect on the gene pool of the race at large. However, it has been suggested that the only path to actually curing such genetically based disorders is by taking advantage of this form of therapy (Annas & Elias, 1992).

The ethical implications of both somatic gene therapy and germ-line gene therapy are comprehensive in scope. Somatic gene therapy has the potential to ease a great amount of pain and suffering caused by ineffective genetic information. Historically humans have supplemented the products of their genes with nutritional decisions, behavioural actions, and medical endeavors. Somatic gene therapy could be the most basic strategy for improving the life and health of individuals in a genetically supplemental manner. If the genetic diseases and disorders can be cured on the level of the genes themselves, then what better way to serve medical purposes? Such a therapy simply accomplishes the same end by a different, more direct means.

Yet, there are ethical problems with somatic gene therapy that accompany the possible benefits previously outlined. Such problems arise in light of the fact that there are different ways in which somatic gene therapy can be applied. Obviously, scientific and medical research ought to benefit those with genetic illnesses. The more important question then arises: what constitutes an illness? Many people are plagued by Tay-Sachs disease, SCID, and cystic fibrosis. What of the people that are plagued by high cholesterol, obesity, and dwarfism? Some people could possibly decide that cosmetic shortcomings, facial asymmetry, and other superficial phenotypes ought to be included under the heading of "bad genes." With such an undefined philosophy of what constitutes genetic illness, one risks falling prey to the ethical anathema of eugenics (Berger et al., 1996).

A semblance of distinction has been made in regards to such ethical conundrums. Negative gene therapy is defined as those genetic alterations that are intended to resolve health problems. Those genetic alterations that are solely intended to improve a healthy individual's genome are defined as positive gene therapy (Berger et al., 1996). The differences between these two applications of gene therapy are significant indeed. Genetic improvement of characteristics that are not directly related to one's health or necessary for survival would lead to the eventual manufacturing of phenotypic traits. Such a lack of control on the genetic market serves no medical purpose, and further has the potential to evolve eugenic tendencies.

Historically, eugenics has repeatedly violated the basic human rights and moral good of humanity as a whole. Such efforts towards racial, or more specifically phenotypic, hygiene have resulted in the death, oppression, and general mistreatment of individuals in every occurrence. Somatic gene therapy that is not medically sanctioned and necessary would be yet another instrument that could be exploited for such ends. As such, our society must be cautious in the application and implementation of such a tool. Somatic gene therapy ought to only be utilized under conditions in which the health of the individual undertaking the procedure will benefit greatly. Gene therapy intended for the purposes of superficial or cosmetic benefits ought not to be carried out. Therefore, as a society, legislation ought to be passed which limits the scope of somatic gene therapy in a reasonable and fair way. Negative somatic gene therapy ought to be encouraged, while positive gene therapy ought to be prohibited.

Finally, the general safety and welfare of the patients receiving somatic gene therapy must be realized and respected. Gene therapy is contemporarily a fledgling science. Researchers and medical physicians must proceed carefully and reasonably with the patient's well-being in mind. There remain health risks that include the spreading of the retrovirus and its genetic contents to tissues other than the tissues that are targeted for therapy (Campell, Tudge, & Wimut, 2000). This risk ought to be professionally assessed and taken into account. As with any other medical procedure, techniques improve with time and experience. In the infancy of gene therapy, precautions must be taken in order to protect the ethical principles of medical endeavors.

The other brand of gene therapy, germ-line therapy contains a distinct collection of ethical and moral concerns. As with somatic gene therapy, germ-line therapy has the possibility of treating and eradicating a great number of genetic plagues. Germ-line therapy is unique in that the affected tissues include the reproductive tissues (Boylan & Brown, 2001). As such, future generations will bear the results of such procedures in their own genetic composition. Therefore, germ-line therapy is a more permanent solution to genetic diseases and disorders because the genes which cause such disorders would be replaced, and the alterations would be preserved in the genome of future generations. Germ-line therapy could be viewed as a more efficient and humane approach to genetic disease treatment (Annas & Elias, 1992). The future generations of unborn humans would be spared the pain and trauma of living with such diseases and possibly undergoing somatic cell treatment themselves. Instead, the genome would have been "cured" in advance, affording future generations an opportunity at a normal existence.

Doubtlessly, the human race would benefit greatly from the ultimate extinction and eradication of devastating genetic diseases and disorders. Yet, here are many ethical dilemmas which result directly from the sheer permanence of germ-line gene therapy. A rash and swift extermination of genetic illnesses permanently could have destructive effects on the evolutionary future of humanity. Human knowledge is shortsighted and quite limited. To think that humans could possibly have the intelligence to predict the evolutionary climate of the future, or bear the responsibility for such alterations in the human population is excessively presumptuous and proud (Annas & Elias, 1992). The reason that sexual reproduction has survived and prospered is that it perpetuates on the concept of diversity. It is through variation that genetics retains a certain degree of plasticity with which to adapt to novel environmental pressures.

For example, sickle cell anemia, some speculate, developed as a medically beneficial phenotype in environmental situations with high rates of malaria infection (Boylan & Brown, 2001). Malaria is a blood born pathogen. Those individuals with sickle cell anemia have a drastically decreased risk of contracting malaria because their blood cells cannot carry the malaria pathogen. However, in environments that do not present a high rate of malaria infection, sickle cell anemia is deleterious (Boylan & Brown, 2001). As evidenced in this situation, environmental pressures directly affect the survival of all living organisms. Furthermore, such environmental pressures often change. The genetic variance of a population helps the species to hedge its bets and survive. If humans were to decide to reduce the variance of the species genome permanently, then those humans ought to know absolutely the environmental conditions that future generations will face. Furthermore, one must note that on an evolutionary scale many genetic products that were once deleterious or neutral have evolved to become neutral or beneficial to an individual (Annas & Elias, 1992). In other words, the genetic disorders that now plague the human race may not be harmful in the future, and might even be advantageous to survival. Therefore, altering the genome of any individual in the germ-line has permanent effects that are not necessarily known to those making the alterations.

There are yet more ethical quandaries that arise in the critical evaluation of germ-line therapy. Firstly, in America and Europe a precedent has been set in legal and ethical medicine stating that the rights of those not able to defend themselves should be defended in cases of medical procedures. The mentally handicapped, mentally ill, young children, and newborns are protected from the imposition of medical treatments that would necessitate consent (Annas & Elias, 1992). In terms of germ-line gene therapy, this ethical policy can be liberally extrapolated to include the genetic composition of future progeny. In other words, germ-line therapy affects directly the health and welfare of individuals not yet conceived. Truly, there are a number of medical treatments that also affect the future of progeny. However, no such treatment affects future children as directly or as singularly as does germ-line gene therapy. As Alexander Capron (1990) has said: The major reason for drawing a line between somatic-cell and germ-line interventions… are that germ-line changes not only run the risk of perpetuating any errors made into future generations of non-consenting "subjects" but also go beyond ordinary medicine and interfere with human evolution. Again, it must be admitted that all of medicine obstructs evolution. But that is inadvertent, whereas with human germ-line genetic engineering, the interference is intentional. (Annas & Elias 1992, p.147)

Also, there remain the clinical risks mentioned previously in the discussion of somatic gene therapy. The clinical risks are heightened to a certain extent however because even less experience and scientifically based knowledge has been gained in the specific field of germ-line therapy (Campell, Tudge, & Wimut, 2000). Finally, there are a number of social concerns that one must consider in due course. Germ-line gene therapy might unfairly burden future generations with facing the problem of genetic discrimination. Historically, racial discrimination (another form of genetic discrimination) has been a repeated offence in many cultures. Time and again, certain ethnicities and cultures have oppressed and stereotyped those of a different background. Germ-line gene therapy offers another avenue of discriminatory possibility based on genetic alterations or the refusal to take such liberties. Also, parents bringing children into the world could feel socially pressured to have their genes and therefore offspring "cured." Financially, insurance companies could potentially require or persuade their clients to receive expensive and extensive genetic treatments in order to avoid the further costs of covering affected children. Such a move to require germ-line gene therapy would make the most sense to insurance companies in order that future profits would be more secure (Annas & Elias, 1992). Finally, and most importantly, the social implication of permanently altering genes in light of past eugenic movements remains a weighty issue. If positive gene therapy and germ-line gene therapy were both allowed to persist, almost certainly the world would see a strong second coming of eugenic ideology (Berger et al., 1996). Why wouldn't parents and nations desire to improve their children by making them the brightest, healthiest, and most advantaged children possible? Those people that lacked the financial capital or influence would be deemed inferior, and subjected to the discrimination of those possessing "superior" genes. Considering the entirety of germ-line gene therapy, such means are difficult to justify. Naturally, the correct application does provide humanity with the substantial benefit in terms of medical practice. However, the ethical implication of correctly and incorrectly applying such methods of medical procedure make the application of germ-line gene therapy, whether positively or negatively applied, impossible to excuse.

Finally, it is important to consider the religious and theological implications of gene therapy in our contemporary understanding of God. Gene therapy by nature transfers the control of genetic traffic from the hand of nature to the hands of humanity. Such a powerful capability must be cautiously approached from a theological standpoint. Many have stated that man, by pushing forward the discovery and employment of genetic technology, has desired to play the role of God. To a certain extent such a statement is valid. Humanity perpetually strives towards the improvement and control of its environment. From taming the wilderness frontiers to scientific pursuits, humanity has proven this desire. Genetics is no exception. Humanity has, and will, continue to attempt improvement in all facets of life. Gene therapy is an especially controversial subject because genetic information is the essence of what makes us human.

When humans unravel and piece together completely the knowledge of genetics, an understanding of the biological processes, functioning, and ultimate humanity will be achieved. Many believe that in such a scientific manner, man will have no place for a higher power and will effectively "play God." Yet, such perceptions are misleading. Through an ethical and religious approach to genetics, man will find the correct uses and misuses of such information (Gehring, 2003). Further, such an understanding of man's basic humanity has the potential not to destroy God, but rather to reinforce the magnificence of biological life. Pope John Paul II has been quoted as saying, "Science can purify religion from error and superstition; religion can purify science from idolatry and false absolutes. Each can draw the other into a wider world, a world in which both can flourish... We need each other to be what we must be, what we are called to be."

Gene therapy in its nascent state has the potential for producing a marked advancement in medical techniques. Simultaneously, its methods have the capacity for terribly unethical and evil application. As a whole, humanity must discern under which conditions and on what terms such technology ought to be put to use. Scientifically, gene therapy has become a reality and will be put to further use in the future. However, the application of this technology has been sparse and limited. As new possibilities are realized, a great deal of ethical and moral decisions must be made. Somatic gene therapy ought to be utilized within negative gene therapy situations. However, when enhancement and other positive gene therapy applications are possible, the benefits of the methods do not compare to the unethical points of contention. Furthermore, germ-line gene therapy, though more efficient and therefore attractive, ought not to be used for the ethical boundaries such methods trespass. Within these ethical considerations, gene therapy will yield a great deal of medical benefit while being cautious not to encroach upon the realm of immorality.

AREA OF RESEARCH IN GENE THERAPY

Gene therapy is a relatively new area of research which in the last ten years has attracted a great deal of attention from scientists, the press, politicians and the public. Unfortunately the initial high expectations have not been fulfilled. Unexpected technical problems, limitations in the technology available and sometimes also insufficient understanding of the pathophysiology of the disorders have been important reasons for this. The general public are somewhat hesitant to embrace this technique in view of the fact that the genetic material of the human subject is deliberately altered. Often this hesitancy is caused by the confusion between *somatic* gene therapy and *germline* gene therapy. With germline gene therapy the genetic material which is passed on to the offspring is altered for a particular purpose. In the Netherlands this type of gene therapy is prohibited. In this statement, gene therapy refers to somatic gene therapy, the technique whereby the genetic makeup of body cells is altered.

The assessment of clinical gene therapy protocols is carried out in the Netherlands by the Central Committee for Research Involving Human Subjects (Centrale Commissie Mensgebonden Onderzoek; CCMO). This is regulated in the *Medical Research Involving Human Subjects Act* (WMO), which fully entered into force on 1 December 1999 [1,2]. In the meantime, the CCMO has received 14 gene therapy protocols for assessment. As a result it has obtained a good impression of current clinical gene therapy research in the Netherlands. In the first 'CCMO year' it has appeared that researchers are unfamiliar as to the role, the procedures of the CCMO and the requirements it applies in assessing gene therapy research proposals. By means of this statement the CCMO wishes to inform researchers about this.

FROM KEMO TO THE CCMO

Before 1 December 1999 clinical gene therapy protocols were reviewed by the Interim Committee on Medical Research Ethics (Kerncommissie Ethiek Medisch Onderzoek; KEMO). During its lifetime this committee reviewed various gene therapy protocols. With the arrival of the CCMO, the KEMO was dissolved on 1 December 1999 [3]. A number of gene therapy protocols which the KEMO was unable to finish testing were passed on to the CCMO on that date. There are important differences between the reviewing of gene therapy protocols by the KEMO and the CCMO. The reviewing of gene therapy protocols by the KEMO was carried out on a voluntary basis, with the KEMO giving an *advice* on the trial protocol reviewed. The CCMO, on the other hand, gives an *assessment*. Without a positive assessment from this committee, the trial cannot start. Unlike the KEMO, the CCMO therefore is not an advisory committee, but is an independent administrative agency (ZBO) which takes decisions. The CCMO is not a part of the Ministry of Health, Welfare and Sport (VWS) and in terms of hierarchy does not come under the Minister of Health, Welfare and Sport. In other words, approval for clinical gene therapy protocols is not given by the Ministry of Health, Welfare and Sport (VWS), but by the independent CCMO.

ASSESSMENT

In carrying out its tasks, the CCMO also adheres to the *General*

Administrative Law Act (Awb). This has consequences for the timescales applied by the CCMO. In the first instance the CCMO has eight weeks to review a protocol. This timescale can be extended on a once only basis to *a reasonable timescale*. For this reasonable timescale the CCMO also applies a period of eight weeks, unless there are urgent reasons for deviating from this. When the review period is extended, the CCMO informs the applicant. For the review timescale, the 'chess clock method' is used. If the CCMO requires further details, the researcher is informed of this. It is then 'his move' and the assessment timescale is interrupted.

The researcher is given a fixed period in which to provide the information which has been requested. The review period continues once the CCMO has received all the additional information. If the information is not provided on time or is incomplete, the CCMO rejects the protocol. A rejection does not prohibit a submission of a new version of the protocol at a later stage. The CCMO will then treat this protocol as a new application. The CCMO endeavours to approach the researcher on one occasion only with the request for additional information, but in practice this is generally not yet feasible. This method has been chosen on the basis of the experience gained with the KEMO. This committee regularly received incomplete research proposals, as a result of which a great deal of time was lost in requesting and reviewing additional information. The CCMO assesses whether the research proposal meets the conditions as they are expressed in Section 3 of the WMO. The CCMO can for instance only reach a positive assessment if it is likely that the research will result in new insights in the field of medicine, the research is carried out by experts and the research cannot be carried out in another way, of a less radical nature (for example by way of animal studies).

THE INFORMATION REQUIRED

The CCMO regards a gene therapy vector as a drug. The consequence of this approach is that the researcher must follow the guidelines for Good Clinical Practice (GCP). A gene therapy vector must satisfy the general quality requirements for a biological drug, such as production in accordance with the rules of Good Manufacturing Practice (GMP). Responsibility for guaranteeing the quality of the vector, in accordance with the Law on the Provision of Medicines (WOG) lies predominantly with the pharmacist of the research institution.

The requirements which the CCMO sets for a protocol are largely in line with those of the European Agency for the Evaluation of Medicinal Products (EMEA) and the American Food and Drug Administration (FDA) [5]. Therefore the researcher has to declare that sufficient preclinical information has been obtained to be able to guarantee the safety of the clinical trial. The preclinical data can be obtained by carrying out *in vitro* studies or *in vivo* animal studies.

The aim of this preclinical phase is to obtain an understanding of the product's activity, the efficacy, the pharmacology, the pharmacokinetics and the toxicity. The preclinical studies also have to provide insight into what would be a safe starting dose, the route of administration, the dosage regimen and any possible toxicity to organs. The scientific basis also forms part of the protocol.

An important safety aspect specific to the use of viral vectors is the testing for the presence of replication-competent viruses. This particularly plays a role if the researcher wants to use vectors derived from retroviruses. In the appendix to this statement you will find additional information about this.

The CCMO attaches great value to carrying out such safety tests prior to the clinical trial. In brief it comes down to the researcher having to demonstrate that it is safe to use the viral vectors in the patient and for the population. In the protocol it must be stated which safety tests are being carried out and how these tests are validated. Internationally this field is still in a state of flux. Recently, for example, the FDA made an addition to the existing guidelines aimed specially at the safety tests for retroviral vectors [7] and the British National Institute for Biological Standards and Control (NIBSC) has recently set up a working party which will handle these issues. The CCMO is involved in this discussion.

SERIOUS ADVERSE EVENTS

In September 1999 an 18 year old boy died in Pennsylvania in the United States during the course of a gene therapy trial [9]. The death of this patient is regarded as the first case of death as a consequence of gene therapy treatment. In the aftermath of this accident, it turned out that in the United States many Serious Adverse Events (SAEs) occurring in gene therapy trials had not been reported or had been reported incompletely to the National Institutes of Health (NIH) [10]. A lack of clarity about the role which this organisation played in recording SAEs in relation to the FDA was probably one of the reasons. The death of the American patient led to much discussion and many articles about the protection of human subjects in clinical (gene therapy) studies [11, 12, 13], the setting-up of the Office for Human Research Protection (OHRP) in the United States [14] and a decrease in the number of gene therapy protocols being submitted to the FDA [15]. On the basis of Section 10, para 1 of the WMO, the CCMO must be informed immediately if 'the research proving to be significantly less favourable to the subject than the research protocol had suggested'. For many researchers it is unclear what the law means by this. In view of the events in the United States and the confusion among gene therapy researchers on this point, the CCMO has decided that for gene therapy research all SAEs must be notified in writing to the CCMO immediately. Part of the SAE reporting procedure includes completing a SAE form which can be obtained from the CCMO-office and the web site.

PRECLINICAL AND CLINICAL GENE THERAPY RESEARCH

The death of the patient in the United States seriously damaged the general public's confidence in gene therapy. In order to restore the general public's confidence, the FDA has recently proposed to make parts of the protocol available to the public [16]. The debate on this public disclosure is still ongoing. Meanwhile there are also positive results to be reported about gene therapy research. In France a number of patients have been treated recently [17]. These patients suffered from a rare X-linked severe immune deficiency (X-SCID) disease which only occurs in young boys. The treatment seems to have been successful in four of the five patients [Cavazzana-Calvo, personal communication], but it is still too early to draw definitive conclusions. The French trial was thoroughly prepared. Before applying the experimental gene therapy to patients, the method was first tested extensively *in vitro* [18] and in a specially developed X-SCID mouse model [19]. Also the first results of gene therapeutic involvement in haemophilia B (factor IX) are promising, although the number of patients 'treated' is small and the *in vivo* production of factor IX is modest [20]. In spite of these interesting initial clinical results, we have to be thoroughly aware of the fact that gene therapy is still at an experimental stage. Great expectations frequently prove to be unrealistic. A great deal of basic knowledge is still needed in the area of (targeted) gene transfer and gene expression regulation. Animal models can make an important contribution in this. Adequate models are still not available for many diseases. It is therefore also very important to develop new animal models. Recent results in this area in cancer research are encouraging [21].

When the preclinical research has yielded sufficient information about the safety and (if possible) the efficacy of the proposed method, the step towards clinical use can be made with caution. The primary responsibility here lies with the researcher. He has to be fully aware of the burden and potential risks for the patients and he has to be able to make a realistic assessment of the benefits of the proposed clinical trial. This is no small task in view of the fact that gene therapy covers a wide area and the risks cannot be fully quantified.

HISTORICAL HIGHLIGHTS OF THE EVOLUTION OF GENE THERAPY

Gene therapy can be described as the introduction of a fully functional and expressible gene into a target cell. This results in permanent curing of the specific genetic disease.

The evolution of gene therapy as a new method is followed the different phases as listed below:

- *Phase I*: Emergence and acceptance of the concept of gene event. Period: 1866-1953
- *Phase II*: Technical Implementation event. Period: 1955–1983

- *Phase III*: Clinical development event. Period: 1975 onwards

On 14/9/1990 after exhaustive reviews by various regulatory panels, the first human gene therapy with severe combined immunodeficiency (SCID), which lacks enzyme adenosine deaminase was done. In this case cloned gene was introduced into lymphocytes (a type of white blood cell) that had been removed from the patient. After culturing the lymphocytes that were not having adenosine deaminase were transfused with the normal gene segment. This cell thus produced the adenosine deaminase enzyme in all the transfused cells. These cells were later returned or transplanted in to the patient.

Severe combined immunodeficiency (SCID), is an inherited genetic disease and in such individuals the blood lymphocytes lack enzyme adenosine deaminase (ADA). This is the key enzyme in purine nucleoside metabolism of the lymphocytes and thus could not synthesize the purines. In the absence of active enzyme adenosine deaminase (ADA), deoxyadenosine is phosphorylated to yield high levels of dATPs that are 50-fold greater than normal levels. This high concentration of dATP inhibits ribonucleotide reductase and thus preventing the synthesis of the other deoxynucleoside triphosphate (dNTPs). This choking off DNA synthesis and thus cell proliferation of the lymphocytes. Thus the lymphocytes concentrations will become lesser and lesser.

This condition leads to the increased fatal infections by the potential pathogens. Since lymphocytes mediate much of the immune response and fatal infections cannot be avoided due to the lack of the fighting lymphocytes special care must be taken and isolated (protective measures should be taken) to avoid all types of potentially fatal infections. The mutations in all eight known ADA variants obtained from SCID patients appear to structurally perturb the active site of adenosine deaminase. As mentioned earlier first breakthrough in gene therapy was carried out at the National Institute of Health by three medical professionals namely Dr. W. French Anderson, Dr. Micheal Blaese, and Dr. Kenneth Culver in 1990, September. They treated a fouryear-old patient Miss. Ashanthi Desilva in the hospital. They took T lymphocytes and transfected with the normal adenosine deaminase gene. Four months later another patient eleven year old Cynthia Cutshall also underwent the similar treatment. Both the patients had been injected with 11 to 12 gene therapy processes. This process effectively treated the disease because even after three years their T lymphocytes showed positive for adenosine deaminase normal gene.

FOUR MAJOR APPROACHES TO GENE THERAPY

In gene therapy different methods are available to correct the defective version of the gene in an individual depending upon how exactly the changes have been made. By using gene therapy approaches large number of diseases are considered as potential source for treatment. There are four major approaches that can be made to obtain the proper function of the gene in the human and they are gene addition, gene replacement, gene switching and gene regulation.

GENE ADDITION

This approach involves addition of a normal gene capable of expressing a normal gene products thus restoring a genetically defective function. The pre-existing mutant (errant) gene is not altered or changed and continue to be present in the system.

GENE REPLACEMENT

In this method mutant gene sequence is altered that has an abnormal gene product. Here the specific excision or repair of a mutant gene is done and transduction by normal gene sequence is carried out. The pre-existing abnormal (errant) gene is no longer present in the cells that are used for the transplantation.

GENE SWITCHING

In this type of gene therapy the establishment of alternative pathways to circumvent mutant gene function is carried out.

GENE REGULATION OR ALTERATION

By this method of gene therapy altering the regulation of normal or mutant gene is carried out. Here the repression or activation of the normal or mutant gene will be done.

TYPES OF GENE THERAPY

In human body mainly two types of cells are present. They are somatic cells and germ cells. These cells are used for correcting the diseased genes to function as normal cells. Based on the types of cells used in the therapy processes it can be two types in humans and they are as follow:

1. Germ line gene therapy
2. Somatic gene therapy

GERM LINE GENE THERAPY

The term indicates the introduction of transgene or transgenic cells into the germ line as well as into the somatic cell population in the next generation. This therapy should have ability to cure patient but some gametes also carry corrected version of the gene for proper phenotype.

Thus, the entire future generation gets cured from the genetic disease. The protocol is relevant for application in humans in the removal of an early embryo with a defective genotype from a pregnant woman and injection with transgenic cells containing the wild type gene.

These cells become part of many tissues of the body, often including the germ line, which will give rise to the gonads. This therapy is more ethically problematic. This approach alters both the somatic cell and germ cell lines of the individual and ensures that the genetic alterations will be passed onto the individual offspring. The ethical issues involved in this approach to gene therapy are currently prohibitive but provide a lively area for debate and discussion about the future of gene therapy. No human germ-line gene therapy has been performed to date.

SOMATIC GENE THERAPY

The aim of the therapy is only to focus on the body (soma) or vegetative cells. This is intended to alter the individual genetic constitution without altering the genetic makeup of the children of that individual. Presently, it is impossible to render an entire body transgenic organism. In such cases it is likely that not all the cells of that tissue need to become transgenic that can ameliorate the overall disease symptoms. The method proceeds by removing some cells from a patient with a defective genotype and making these cells transgenic through the introduction of copies of the cloned wild type gene. The transgenic cells are then reintroduced into the patient's body which in turn providing normal function. In addition, various strategies for implementing somatic cell gene therapy are beginning to emerge and can be grouped under the broad categories of ex-vivo gene therapy, in-vivo gene therapy and antisense therapy. The gene therapy can be done at two different levels of individual.

The first one is embryo gene therapy in this method genetic constitution of embryo is altered after zgyote is formed. The second method is patient therapy, in this healthy gene is introduced into affected person tissues. The healthy gene overcomes the defect without affecting the inheritance. The difference between the germ line therapy and somatic gene therapy are listed.

SOMATIC GENE THERAPY

With the discovery of the molecular basis of DNA transformation in bacteria in which a gene from one strain can be transferred to and expressed in another strain, human genetic diseases might be cured by introduction of normal gene into the appropriate somatic cell type. During the 1980s, the prospects of human somatic gene therapy techniques were developed, eukaryotic expression vectors were generated. With this transgenic experiments with mice became routine practice.

SOURCE OF CELLS USED FOR THE THERAPY FOR GENETIC MANIPULATION

The source of cells used for the therapy for genetic manipulation can either be autologous *i.e.*, from the patient itself or from a compatible donor. The autologous source would naturally avoid the problems that may arise with adverse immune responses.

The compatible donor may be relative or siblings that is generally identified by the tissue typing procedures. Different cell types that are currently used for ex-vivo gene therapy are the totipotent embryonic blood stem cells from the bone marrow, hepatocytes, vascular smooth muscle cells, fibroblasts, pulmonary epithelial cells and tumor infiltrating lymphocytes. The totipotent bone marrow cells are a popular choice for gene therapy because these will differentiate into a number of different cell types in mature blood including B-lymphocytes and T-lymphocytes, macrophages, red blood cells, platelets and osteoclasts. The ex-vivo gene therapy approach involves the following major steps.

- Collect the cells from the affected patient and culture the cells that are to be manipulated.
- Transfer the desired gene in to the cultured cells by using recombinant DNA technology.
- Select for the cells that have been genetically transformed or genetically corrected (remedial cells) and expand the cell numbers in culture.
- Either infuse or transplant these cells back into the patients.
- Observe for the functions of introduced cells in the tissues or organs.

The gene transfer technology for ex-vivo gene therapy should be efficient and the remedial gene should be both stably maintained and persistently expressed in the transplanted or infused persons. The various vectors are used for such gene transfer protocols. These vectors can be classified as viral vectors and the non-viral vectors. The viral vectors are generally retroviral vectors, adenoviral vectors, adeno-associated viruses and herpes viral vectors. The non-viral vectors include naked DNA microinjection, gene gun method and liposomes.

Large-scale production of bio-pharmaceuticals normally utilises "transformed" (or continuous) cell lines. These grow faster than ordinary cells and can be grown at greater densities, so they give higher yields in bioreactors. They are also easier to maintain in simple culture media. Like cancer cells, "transformed" cell lines do not stop dividing to repair any damage, instead, they keep dividing as long as there are sufficient nutrients and that other suitable chemical and physical conditions for survival are maintained.

In animal tissues and organs, a balance is achieved between and within cell populations. But in cell culture, this is not the case: cells therefore continue to divide forming a physiological and morphological uniform population until death is triggered. Significant progress has been made in recent years in the development of process strategies for animal cells that provide optimal growth conditions and maximise production. Our growing understanding of the molecular mechanisms that regulate cell proliferation, cell death, and the synthesis and secretion of protein products has been increasingly exploited for effective design and operation of simple and intensive production processes. Nevertheless, the scale up and optimisation of animal cell cultures to meet production demands has been problematic, due to the low productivity and instability of the cell lines used.

We have separated cell proliferation from cell death in order to find a way of keeping cells in the non-dividing and non-dying state in which they are most productive. We are investigating ways of maintaining high antibody production in myeloma and Chinese Hamster Ovary cell lines.

The approach being used involves the direct modification of the metabolic and regulatory networks in cells, sometimes called metabolic engineering. Their strategy is to use recombinant DNA techniques to enhance cell survival and control proliferation. Control over cell proliferation and cell death would significantly improve productivity, and would allow the operators to integrate process operations of high predictability and reliability.

The death of cells invariably takes place by a process called apoptosis. This is a programmed physiological process by which cells commit suicide in response to damage or the presence or absence of specific environmental stimuli, for example, depletion of nutrients. The Birmingham scientists have already characterised several stresses that trigger apoptosis in large-scale bioreactors.

We have also shown that ectopic expression of the anti-apoptotic cellular protein known as Bcl-2 inhibits apoptosis in NS0 myeloma, CHO and hybridoma cell lines. Bcl-2 can promote survival at very low dilution rate in continuous culture. In general, the group has found that by introducing anti-apoptotic genes into cells they can increase the robustness and survival of cells in culture. This shows that we can enhance the survival and productivity of cell lines by manipulating their "death pathway.

This opens up interesting new directions in biochemical engineering. It was also possible to control cell proliferation by using our knowledge of how cell growth and the cell division cycle are regulated by an extended family of enzymes called cyclin-dependent kinases, or cdks.

The activity of cdks, and progress through the cell cycle, depends not only on the availability of cyclin regulatory subunits but also the levels of specific cdk inhibitory proteins (ckis). Two members of the KIP family of ckis, p21CIP1 and p27KIP1, inhibit the activity of the G1-phase cyclin E dependent kinase cdk2, and are normally involved in causing cell cycle arrest, for example in cells containing damaged DNA or in the presence of TGF-beta.

We have introduced an inducible copy of the p21CIP1 gene into cell lines to produce anchorage-independent, proliferation-controlled cell lines, which arrest in G1-phase of the cell cycle with enhanced secreted antibody production.

The improved productivity is independent of the inducible promoter used to achieve this cytostasis (the non-dividing state).

We have extended the survival period of these cells by constitutively over-expressing Bcl-2 protein. This has inhibited apoptosis while maintaining the high productivity associated with cytostasis. We are now carrying out experiments to understand how cytostatic arrest increases productivity. We are employing genomics and metabolomics techniques to identify the genes that are involved in antibody synthesis and cell size; two main processes associated with the over-expression

ARTICULAR CARTILAGE

Damaged articular cartilage has a limited ability to repair itself due to the absence of vascularization and nerve endings in the tissue. Traditional therapies to repair damaged cartilage include alloplastic and allogenic implants and more recently autologous chondrocyte transplantation. The former therapies are limited by donor site morbidity, while the latter therapy requires surgical removal of healthy cartilage and is limited by the size of the defect. As an alternative to these current therapies, efforts in tissue engineering of cartilage have led to the development of biocompatible, biodegradable scaffolds onto which chondrocytes are seeded. One of the challenges that tissue engineers will have to address in the near future is the development of feasible large-scale cell expansion processes.

The estimated number of articular cartilage incidences worldwide is around 30 million cases of knee osteoarthritis and 1.2 million cases of focal defects every year 8. If tissue engineering products are to be used for the treatment of these incidences, it is crucial to estimate the scale of cell production needed for all these repair procedures, but such essential question that is rarely approached in reported tissue engineering studies. The table below gives an idea of the dimensional aspect of articular cartilage repair and the estimated cell necessities associated. Routine tissue culturing methodologies can hardly cope with the scale of cell production required for these procedures. Expansion of cell population in vitro has become an essential step in the process of tissue engineering of articular cartilage, and optimization of the culture conditions and expansion protocols are fundamental issues that need to be addressed.

Currently our aim is to tissue engineer cartilage that can be used for biological repair of damaged articular cartilage or degenerative joint disease. In order to do this cartilage will be engineered that is more similar to natural tissue than current in vitro engineered cartilage. An important issue to address is that of nutrient diffusion limited thickness of engineered cartilage. We propose to develop a novel bioreactor to culture tissue engineered cartilage constructs under hydrodynamic loading and to test the hypothesis that mechanical stimulation to tissue specimens mediates the effectiveness of 3-D scaffolds for cartilage tissue engineering.

Cartilage in human joints is exposed continually to mechanical loading, which is considered to be essential for its homeostasis, and for achieving functional tissue engineered cartilage repair in 3-D biodegradable scaffolds. The cultivation of chondrocytes in polymer scaffolds leads to implants that may potentially be used to repair damaged joint cartilage or for reconstructive surgery. For this technique to be medically applicable the physical parameters that govern cell growth and differentiation in the polymer scaffold must be understood and optimised.

Cultured cells from different mammals can be fused to produce interspecific hybrids, which have been widely used in somatic-cell genetics. For instance, hybrids can be prepared from human cells and mutant mouse cells that lack an enzyme required for synthesis of a particular essential metabolite.

As the human-mouse hybrid cells grow and divide, they gradually lose human chromosomes in random order, but retain the mouse chromosomes. In a medium that can support growth of both the human cells and mutant mouse cells, the hybrids eventually lose all human chromosomes.

However, in a medium lacking the essential metabolite that the mouse cells cannot produce, the one human chromosome that contains the gene encoding the needed enzyme will be retained, because any hybrid cells that lose it following mitosis will die. All other human chromosomes eventually are lost.

By using different mutant mouse cells and media in which they cannot grow, researchers have prepared various panels of hybrid cell lines. Each cell line in a panel contains either a single human chromosome or a small number of human chromosomes, and a full set of mouse chromosomes. Because each chromosome can be identified visually under a light microscope, such hybrid cells provide a means for assigning, or "mapping," individual genes to specific chromosomes. For example, suppose a hybrid cell line is shown microscopically to contain a particular human chromosome.

That hybrid cell line can then be tested biochemically for the presence of various human enzymes, exposed to specific antibodies to detect human surface antigens, or subjected to DNA hybridization and cloning techniques to locate particular human DNA sequences. The genes encoding a human protein or containing a human DNA sequence detected in such tests must be located on the particular human chromosome carried by the cell line being tested. Panels of hybrids between normal mouse and mutant hamster cells also have been established; in these hybrid cells, the majority of mouse chromosomes are lost, allowing mouse genes to be mapped to specific mouse chromosomes

A number of mutant cell lines lacking the enzyme needed to catalyze one of the steps in a salvage pathway have been isolated. For example, cell lines lacking thymidine kinase (TK) can be selected because such cells are resistant to the otherwise toxic thymidine analog 5-bromodeoxyuridine. Cells containing TK convert 5-bromodeoxyuridine into 5-bromodeoxyuridine monophosphate. This nucleoside mono- phosphate is then converted into a nucleoside triphosphate by other enzymes and is incorporated by DNA polymerase into DNA, where it exerts its toxic effects.

This pathway is blocked in cells with a *TK* mutation that prevents production of functional TK enzyme. Hence, TK" mutants are resis- tant to the toxic effects of 5-bromodeoxyuridine. Similarly, cells lacking the HGPRT enzyme have been selected because they are resistant to the otherwise toxic guanine analog 6-thioguanine. As we will see next, HGPRT" cells and TK" cells are useful partners in cell fusions with one another or with cells that have salvage-pathway enzymes but that are differentiated and cannot grow in culture by themselves.

The medium most often used to select hybrid cells is called *HAT medium,* because it contains hypoxanthine (a purine), aminopterin, and thymidine. Normal cells can grow in HAT medium because even though aminopterin blocks de novo synthesis of purines and TMP, the thymidine in the media is transported into the cell and converted to TMP by TK and the hypoxanthine is transported and converted into usable purines by HGPRT. On the other hand, neither TK" nor HGPRT" cells can grow in HAT medium because each lacks an enzyme of the salvage pathway. However, hybrids formed by fusion of these two mutants will carry a normal *TK* gene from the HGPRT" parent and a normal *HGPRT* gene from the TK"parent. The hybrids thus will produce both functional salvage-pathway enzymes and grow on HAT medium. Likewise, hybrids formed by fusion of mutant cells and normal cells can grow in HAT medium.

For many types of studies involving antibodies, monoclonal antibody is preferable to polyclonal antibody. However, biochemical purification of monoclonal antibody from serum is not feasible, in part because the concentration of any given antibody is quite low. For this reason, researchers looked to culture techniques in order to obtain usable quantities of monoclonal antibody. Because primary cultures of normal B lymphocytes do not grow indefinitely, such cultures have limited usefulness for production of monoclonal antibody. This limitation can be avoided by fusing normal B lymphocytes with oncogenically transformed lymphocytes called*myeloma cells,* which are immortal.

Fusion of a myeloma cell with a normal antibody-producing cell from a rat or mouse spleen yields a hybrid that proliferates into a clone called a hybridoma. Like myeloma cells, hybridoma cells are immortal. Each hybridoma produces the monoclonal antibody encoded by its B-lymphocyte partner. Many different myeloma cell lines from mice and rats have been established; from these, HGPRT" lines have been selected based on their resistance to 6-thioguanine as described above. If such mutant myeloma cells are fused with normal B lymphocytes, any fused cells that result can grow in HAT medium, but the parental cells cannot. Each selected hybridoma then is tested for production of the desired antibody; any clone producing that antibody then is grown in large cultures, from which a substantial quantity of pure monoclonal antibody can be obtained.

Prior to cell counting, one should have available a shake flask culture (of about 400mL) with at least 10^6 cells. A shake flask culture with approximately 10^6 cells is obtained by allowing a shake flask culture with roughly 0.5 million viable cells to shake in a CO_2incubator (34 degrees C) for 36 to 48 hours. After the shake flask is removed from the incubator, 5mL of the culture must be transferred to a 15 mL centrifuge tube with standard sterile operating procedure.

After the lid is screwed back on the lid of the cell culture flask, 5mL of 1x trypsin-EDTA solution must then be transferred by pipette into the same 15 mL centrifuge tube. The trypsin solution serves to detach the "clumps" of animal cells. The trypsinized solution must then be rocked in the CO_2 incubator for approximately 15 minutes. To ensure that the more resistant CHO cell clumps are detached, the trypsinized sample should then be pipetted up and down several times; the tip of the pipette should serve to shear the more resistant clumps. The trypsinized sample is then ready to be stained with 0.4 per cent (w/v) trypan blue. A 100-μL sample of the trypsinized solution should be transferred into a 1.5-mL centrifuge tube. A 100-μL sample of the trypan blue should then be added to the trypsinized animal cell culture sample. This mixture should be quickly mixed to give the trypan blue a chance to stain the non-viable cells (a mini-vortex machine may be used). A 20-μL sample of this trypan-stained mixture should then be transferred to the edge of a cover-slipped hemocytometer. After this sample is transferred, the hemocytometer should be placed under the microscope (at 10x) for cell counting.

5

Animals Blood Circulation

BLOOD OF ANIMALS

When Animal Blood is used in cooking: for example, in thesausagesknown as blackpuddings, the addition of several aromaticspicesis necessary so as to overcome its alkaline flatness, and lack of savour. "Blood," says Dr. Thudicum, "is not capable of giving a savoury extract (to gravy), although the blood of each species of eatable animal has its particular, and distinctive flavour; that of the ox, and cow being remarkably redolent of musk." But among civilized nations the pig is the only animal of which the Blood furnishes a distinct article of food; mixed withfat, and spices, whilst enclosed in prepared intestines, this pig's blood is made into black puddings.

Chemically the Blood of animals contains a considerable quantity of iron, besides albumin, fibrin, hydrogen, some traces of prussicacid, and some empyreumaticoil. The serum, or thin part of the Blood, includes sulphur. Experimentally it turns out that the blood of snails, which is colourless, contains as much iron as that of the ox, or calf, this fact going to prove that the red colour of animal Blood is not due, as is generally supposed, to the presence of iron in that fluid. The saline constituents of Blood are phosphates of lime, and magnesium, with chlorides, sulphates, and phosphates of potash and soda.

In Pickwick, Mr. Roker, the coarse turnkey at the Fleet Prison for debt, when showing Mr. Pickwick what were to be his wretched quarters there, turned fiercely round on him whilst he was mildly expostulating, and uttered in an excited fashion "certain unpleasant invocations concerning his own eyes, limbs, and circulating fluids".

Pliny tells us that the Blood of animals (and, indeed, human Blood as well) was administered in his time for curative purposes; so likewise the Blood of the ox is in medicinal vogue today in certain parts of the Western Hemisphere. This is because of the well-ascertained fact that iron, particularly its organicsalt (haemoglobin) as found in Blood, forms one of the most important constituents. It may be thus supplied from the pig in the culinary form of black puddings; as likewise from the ox, or sheep, if so desired. Among the Boers in South Africa dog's Blood is an established remedy for convulsions, and fits. It is of modern discovery that in health the humanliverhas to receive a comparatively large allowance of iron, for carrying on the vital processes of combustion and oxidation, as its special functions. This iron is best obtained from the food, and not through any form of physics. We know that many animals, especially beasts of prey, derive their needful supply of iron exclusively from meat containing a large proportion of Blood, which is rich in organic iron.

Towards overcoming the natural repugnance of a patient to drinking animal Blood for acquiring its iron remedially, some skilful foreign chemists have produced this essential product of late in a compact form, which they term "Sanguinal," as a brownish red powder consisting (as is asserted) of pure crystallised haemoglobin, with the mineral Blood constituents, and of muscle albumin. Hypothetically it is fair to suppose that in this way the red corpuscles of a bloodless patient may be beneficially augmented.

Pepys observed about a Mr. Andrews who was dining with him, "What an odd, strange fancy he hath to raw meat, that he eats it with no pleasure unless the Blood run about his chops," which it did now by a leg of mutton that was not above half-boiled; but "it seems at home all his meat is dressed so, beefand all".

Practical experiments have shown that metallic iron, in whatever form it is administered medicinally, can be recovered from the excretions, absolutely undiminished in quantity, so that evidently no particle thereof is assimilated into the system. Nevertheless, the machinery of red Blood-making is undoubtedly started afresh by giving iron, whether in food, or in physic (much more problematically). In 1902 Professor Bunge read an important paper on "Iron in Medicine" before the German Medical Congress.

He advocated an increased attention to foods containing iron, as a substitute for its administration in drug-form. "Spinach," said he, "is richer in iron than yolk of egg, and yolk of egg than beef; milk is almost devoid of iron; and, as if to provide against this defect, the Blood of the infant mammal is more plentifully endowed with the essential ingredients than that of adults, thus showing that nature is always self-provident." Garden spinach (one of the "Goosefoot" order), than which no better blood-purifier grows amongst vegetables, contains iron as one of its most abundant salts; hence it is a valuable food for bloodless persons; moreover, in both salinity, and digestibility it leads the kitchen greens, its amount of salts being 2 per cent, whereby it helps to furnish red colouring matter (haemoglobin) for the blood.

In the fruit world even the apple does not afford so much iron as this vegetable, neither does the strawberry. Spinach insists on having a rich soil in which to grow, out of which it extracts a large proportion of saline matters. Its full green juice abounds in chlorophyll, insomuch that the spinach may be cooked entirely in its own fluids, and in the steam which will arise from them. This brilliant green principle of colour, elaborated from the yellow and blue rays of the sunlight, is peculiarly salubrious. Evelyn (Acetaria) has said, "Spinach being boil'd to a pulp, and without other water than its own moisture, is a most excellent condiment for almost all sorts of boil'd flesh, and may accompany a sick man's diet. 'Tis laxative and emollient, and therefore profitable for the aged".

Savoy, a nutritious, and wholesome companion of spinach, contains the greatest amount of vegetable oil of all this class of kitchen plants; and spinach runs the luxuriant Savoy very close in its complement of bland oil-salts, which render the juices nourishing. Quite half a pint of spinach-oil might be expressed from a hundred pounds of the vegetable, and sometimes more than this from the same quantity of Savoy.

BLOOD COMPONENTS OF ANIMAL

Whole blood is collected from a donor animal for blood transfusion purposes into a blood bag containing citrate phosphate dextrose as the anticoagulant. Donor animals should be selected with care and strict attention should be paid to the blood collection technique to maintain sterility at all times.

Once the blood has been collected, it can be kept and used in its natural state or can be converted into a variety of components. These blood components are packed red blood cells, platelet-rich plasma, platelet concentrates, fresh plasma, fresh frozen plasma, frozen plasma, cryoprecipitate and cryosupernant.

Double, triple and quadruple blood bags (a single whole blood collection bag with various satellite bags) are used for producing and separating components within a sterile closed environment. Blood component production is highly desirable for several reasons, including maximization of the yield of products from a single blood donation and ability to use the optimal products (in high concentrations) for specific diseases (thus minimizing the amount of unnecessary foreign material the recipient animal is exposed to). The most commonly available blood components are whole blood, packed red blood cells, fresh frozen plasma and cryoprecipitate.

Some definitions may be useful for understanding blood components:

- *Whole blood:* This is blood collected directly from a donor animal into a blood transfusion bag containing citrate-phosphate-dextrose with (CPDA-1) or without adenine (CPD) as an anticoagulant. In general, for dogs, a 500 ml transfusion bag is used (which contains approximately 63 ml of anticoagulant, obtaining approximately 450 ml blood from the donor dog).

  Whole blood can be used immediately or stored at 4 C for future transfusion or separation into blood components. Additives can be used to improve red cell storage viability.
- *Packed red blood cells:* These are a concentrated source of red blood

cells that remain in a small amount of plasma upon removal of supernatant plasma after centrifugation of the blood transfusion bag containing whole blood.

Packed cells are an excellent source of red blood cells for anemic animals that need the additional oxygen-carrying capacity these cells provide.

Packed red cells can be used immediately or stored at 4°C. They need dilution in sterile isotonic solution prior to infusion (due to the high hematocrit, packed cells without dilution are very thick and flow sluggishly through infusion lines). The current standard for red cell storage is that there should be about 75 per cent posttransfusion viability after storage, that is at least 75 per cent of the transfused red cells must remain in the recipient's circulation 24 hours after transfusion. The posttransfusion viability of canine red cells collected into CPDA-1 is 20 days.

Additives have been developed that can be added in a sterile fashion (they are attached to the original transfusion bag as a satellite bag) to packed red cells (or whole blood bags) to optimise red cell storage viability. These additives usually contain dextrose and adenine for red cell energy metabolism and mannitol, which decreases red cell lysis. Several additives have been evaluated with canine red blood cells. Both Nutricel and Adsol increase the posttransfusion viability of stored canine red cells to 35 and 37 days, respectively.

- *Platelet-rich plasma (PRP):* Platelet-rich plasma is produced by separating plasma from red blood cells (within 6 hours of whole blood collection) using a slow spin to prevent pelleting of the platelets. Platelet-rich plasma is useful for the treatment of disorders of platelet number and function but must be infused within 8 to 12 hours (maintained at room temperature) due to platelet instability (which limits its use).
- *Platelet concentrates:* Platelet concentrates are produced by separating most of the plasma from platelet-rich plasma. Platelet concentrates are used similarly to platelet-rich plasma.
- *Platelet-poor plasma (PPP):* Platelet-poor plasma is produced by separating plasma from red blood cells using a high spin to pellet platelets with the red cells. This is the starting point for production of most blood components. Separation of platelets is desirable because they provide an additional source of foreign antigens and micro-aggregation can cause transfusion reactions. Approximately 200 to 400 ml of plasma are obtained from each whole blood transfusion bag. One unit of plasma is defined as that obtained from a single whole blood transfusion bag.

- *Fresh plasma:* This is platelet-poor plasma that is separated from red cells and infused within 6 hours of blood collection (with the blood bag being maintained at 4°C until separation).

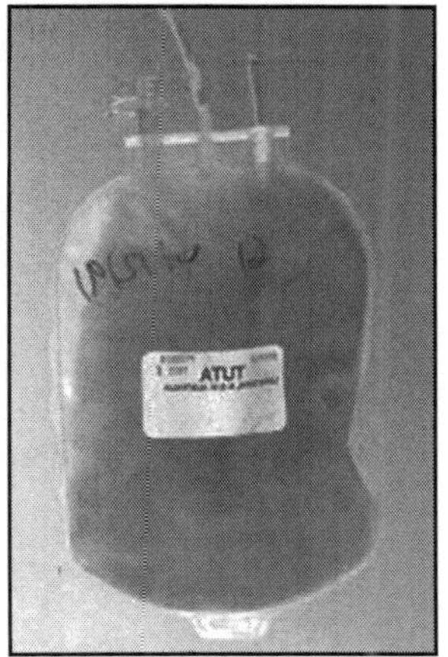

- *Fresh frozen plasma (FFP):* This is platelet-poor plasma that is separated from red cells within 6 hours of blood collection and frozen in a dedicated freezer (at or below –20°C). Fresh frozen plasma and fresh plasma contain all coagulation factors and plasma proteins (such as albumin).

  Fresh frozen plasma is stable for 1 year if maintained in a dedicated freezer (one that does not undergo freeze-thaw cycles like household freezers) at or below –20 C.

  One unit of FFP is defined as that obtained from a single whole blood transfusion bag (approximately 250 ml).
- *Cryoprecipitate (CPP):* Cryoprecipitate is a concentrated source of von Willebrand factor, fibrinogen, factor VIII and fibronectin. It is produced by slow-thawing (at 4°C) fresh frozen plasma, followed by centrifugation at 4°C. Cryoprecipitable proteins (those mentioned above) precipitate at this temperature and are maintained in a very small amount of remaining plasma (approximately 1/10 of the starting volume of FFP, or about 20 to 30 ml). Cryoprecipitate is stable for 1 year from the date of collection of the whole blood for transfusion purposes (not the date of preparation of the product) if maintained at or below–20 C in a dedicated freezer. One unit of cryoprecipitate is defined as that obtained from a single FFP bag (approximately 250 mL plasma).
- *Cryosupernant (Cryosuper):* Cryosupernant is the remaining plasma after removal of the pelleted cryoprecipitate. This is a source of all coagulation and plasma proteins, except for factor VIII, fibrinogen, von Willebrand factor and fibronectin. Cryosupernant is stable for 5 years if stored in a dedicated freezer at or below –20°C.

- *Frozen plasma:* Frozen plasma has several sources. It can be plasma that is separated and frozen longer than 6 hours after whole blood collection. It can be fresh plasma that is not used within 6 hours of collection and then frozen. It can be fresh frozen plasma maintained for longer than 1 year in a dedicated freezer. Frozen plasma lacks certain coagulation factors which are quite unstable.

BLOOD COAGULATION

ANTICOAGULANTS

Heparin is the prototypical anticoagulant. It was discovered in the liver of dogs in 1916 and was purified in the 1930s. It requires a plasma factor called antithrombin for its anticoagulant action. Heparin can be purified from the tissues of different animals. It cannot be given orally. It is widely used in medical practice for a number of reasons and clinical situations. It is commonly used to keep IV lines from clotting off, and it is used in a number of clinical situations where either a clot has occurred or there is a high risk of a clot occurring and causing serious medical complications. Its effect is immediate but has a short duration of activity (half-life) and therefore is most often given continuously through an IV line.

Alternatively, it is sometimes given as subcutaneous (under the skin) injections on a regular and frequent schedule (such as every four to six hours). The most feared side effects of heparin are bleeding and *heparin induced thrombocytopenia* (H.I.T.) Heparin administration must be monitored closely by means of a test called the P.T.T. (partial thromboplastin time). Other side effects of heparin include osteoporosis (with long term use), allergic reactions and elevation of liver enzymes. Warfarin is the dominant oral anticoagulant in the United States. It is derived from dicumarol, which is found in moldy sweet clover and caused an epidemic of cattle deaths in the early 1920s before the cause was discovered. In the 1930s it was discovered that vitamin K is necessary to prevent bleeding, and dicumarol and related compounds such as warfarin are known to work by inhibiting coagulation factors that require vitamin K (but this fact was not discovered until the 1970s). Warfarin received approval in the 1950s for clinical use.

Side effects of warfarin include paradoxically increased clotting activity for the first 24 to 36 hours of treatment, allergic reaction, drug interactions and severe skin reaction (necrosis) that may be associated with other conditions. The anticoagulation effect of warfarin is delayed for several days before becoming apparent. Therefore, heparin is often started in the hospital until the warfarin effect occurs.

Low Molecular Weight Heparins are stripped down versions of conventional (sometimes also called unfractionated.) heparin. Low molecular weight heparin molecules are about one third the size of unfractionated heparin, but contain the key five sugar sequence (pentasaccharide) that is responsible for the anticoagulant activity. Low molecular weight heparins can be given subcutaneously in a fixed dose and do not require monitoring. They can be used at home.

Fondaparinux is the most stripped down version of heparin. It consists of the five sugar sequence (pentasaccharide) that gives unfractionated heparin its anticoagulant activity. It is manufactured synthetically. Bivalirudin is another synthetic anticoagulant. It is derived from the anti-clotting substance found in leeches.

New Types of Oral Anticoagulants

The ideal anticoagulant drug would be one with a wide therapeutic index, oral administration, once a day administration, easy reversibility and availability of an effective antidote, short half life, lack of need for laboratory monitoring and reasonable cost.

While unfractionated heparin, low molecular weight heparins and warfarin in their own respects fulfil some of the above characteristics, none of them are by any means ideal and they all fall far short of fulfilling all of the above criteria.

- *Direct thrombin inhibitors (DTI): dabigatran* (Pradaxa-Boehringer Ingelheim) is a DTI that has been shown to be as effective and safe as enoxaparin (Lovenox) in reducing VTE after orthopedic surgery. As the name implies, it works by directly inhibiting the thombin molecule in the coagulation cascade. It does not require an intermediate molecule to be effective. It received approval in Europe for use after orthopedic operations to prevent blood clots from occurring.
- *Factor Xa inhibitors:* These drugs work by inhibiting the molecule Factor Xa in the coagulation cascade. Rivaroxaban (Xarelto-Bayer) has received approval in Canada and the EU for prevention of VTE in patients undergoing elective total hip or total knee replacement surgery. Apixaban failed to meet pre-specified criteria of non-inferiority in a study comparing it to enoxaparin in post-op patients but was numerically very similar in efficacy and statistically superior in terms of bleeding complications.

WHAT IS COAGULATION

Coagulation is the series of events that result in the formation of a clot. In the body, coagulation occurs after any injury to a blood vessel or tissue, in order to stop bleeding. Coagulation involves the interaction of cells lining the injured blood vessel (*endothelial cells*), Specialised blood cells called *platelets* that form a plug in the region of the damaged blood vessel, and circulating *coagulation factors*.

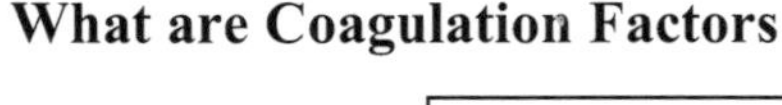

What are Coagulation Factors

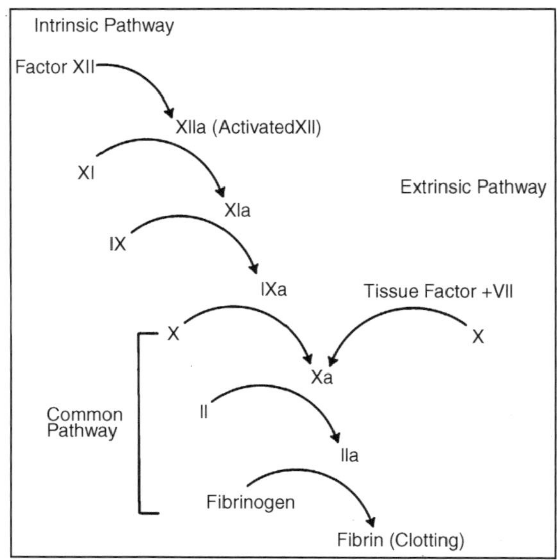

Coagulation factors are substances that are made in the liver and circulate in the blood stream. They become 'activated' when blood vessel or tissue injury occurs, and in co-operation with platelets, produce a clot at the site of injury. There are thirteen different coagulation factors. These factors are activated in a specific sequence, following two different pathways or sequences (the intrinsic and extrinsic pathways), and a final common pathway. In the body, there is interaction between the intrinsic and extrinsic coagulation pathways. In the laboratory, the activity of the coagulation factors comprising these pathways is measured separately.

When are Coagulation Tests Performed

Coagulation tests may be undertaken for a variety of reasons. Some breeds of dogs are known to have a higher incidence of clotting factor deficiencies and with these breeds, coagulation tests may be used for 'screening purposes' prior to diagnostic or surgical procedures.

"Some breeds of dogs are known to have a higher incidence of clotting factor deficiencies..." If an animal experiences episodes of bleeding for an unknown reason, evaluation of clotting function should be undertaken. Severe liver disease may lead to problems with blood clotting because the liver is the site of production of coagulation factors.

> Are there any screening tests that can be used in the veterinary clinic to evaluate coagulation?
>
> Yes. The most commonly performed test is the evaluation of platelet numbers. Platelets are blood components that are involved in the formation of platelet 'plugs' at the site of blood vessel injury. Platelets

not only provide the initial 'patch' at injured sites, they also form a framework that allows coagulation factors to accumulate at the injured site and subsequently form a more permanent fibrin clot.

Platelet numbers can be determined readily, and only a single blood sample is needed. Platelets are usually evaluated as part of a routine complete blood count (CBC) that also evaluates the red blood cell and white blood cell components of blood. A marked decrease in total platelet numbers (termed *thrombocytopenia*) may result in episodes of bleeding. There are many causes of *thrombocytopenia,* including immune-mediated disease, bone marrow disorders, and infectious causes, among others.

"...*thrombocytopenia* may result in episodes of bleeding."

While changes in the red and white blood cells on a CBC do not provide specific information about coagulation, they may alert your veterinarian to investigate further. A reduction in red cell numbers *(anemia)* may occur as a consequence of coagulation factor or platelet deficiency, while changes in the white blood cell numbers or types may indicate an underlying inflammatory condition that may be responsible for the *thrombocytopenia*. Because clotting factors are made in the liver, and because many diseases can result in thrombocytopenia, a serum biochemistry panel that evaluates liver function may be helpful.

The buccal mucosal bleeding time (BMBT) is a test which can be performed in the clinic and which evaluates the ability of platelets to form a platelet plug. A spring-loaded cassette is used to produce a small, precise cut on the inside of the upper lip, and the length of time required for bleeding to stop is measured. This time may be lengthened by a decreased platelet count or by decreased platelet function. The BMBT is usually reserved for patients that have normal platelet numbers but have questionable platelet function.

The ACT (activated clotting time) test can be performed in the veterinary clinic. To perform this test a blood sample is added to a special ACT tube containing a substance (usually diatomaceous earth) that activates the coagulation factors in both the intrinsic and common pathways. A severe decrease (less than 5 per cent of normal activity) in these clotting factors or a severe decrease in platelet numbers will lead to an increase in the length of time it takes the blood to clot after it has been added to the ACT tube. The ACT is not very sensitive to milder decreases in clotting factors.

What specific tests are used to measure the activity of coagulation factors?

"More specific coagulation tests are sent to a veterinary or hematological (blood) reference laboratory..." These more specific coagulation tests are sent to a veterinary or hematological (blood) reference laboratory and require a single blood sample.

The PTT (activated partial thromboplastin time) measures the functional activity of the coagulation factors involved in the *intrinsic* and common coagulation pathways. The PTT measures the time that it takes for a fibrin clot to form after calcium and an activator are added to a blood sample. The results of the PTT will be increased (that is, the time required for a clot to form will increase) when there is at least a 70 per cent decrease in the activity of a single factor in either of the intrinsic or common coagulation pathways. Smaller decreases in multiple factors may also result in an increase in the PTT. Anti-coagulant therapy such as heparin will increase the PTT.

The PT (prothrombin time) measures the functional activity of the coagulation factors involved in the *extrinsic* and common pathways. The PT does this by measuring the amount of time it takes for a fibrin clot to form after calcium and a tissue-activating factor are added to a blood sample. The results of the PT will be increased (that is, the time required for a clot to form will increase) when there is at least a 70 per cent decrease in the activity of a single factor in either of the extrinsic or common pathways. Because of the specific factors involved in the extrinsic coagulation pathways, an increase in PTT is often noted with conditions such as vitamin K deficiency and rodenticide (rat poison) toxicity.

"...Increase in PTT is often noted with conditions such as vitamin K deficiency and rodenticide (rat poison) toxicity."

The TT (thrombin time) measures the functional activity of the *common* coagulation pathway, and more specifically of *fibrinogen*. The conversion of fibrinogen to a fibrin clot is the end point of both the intrinsic and extrinsic pathways in the body. Both pathways converge to the final common pathway that results in fibrin clot formation. The TT measures the amount of time it takes for a fibrin clot to form after calcium and thrombin (a coagulation factor) is added to a blood sample. Since the TT is dependent upon both the amount and the normal functional activity of fibrinogen, any condition that depletes the amount of fibrinogen or makes it unable to function properly will cause an increase in TT. Examples of such conditions include hereditary deficiencies in the amount or functional activity of fibrinogen and increased use/consumption of fibrinogen in the body because of excessive blood coagulation (clotting). Such excessive coagulation can occur because of shock, severe infections, and tumours, to name a few examples.

Von Willebrand's factor testing at a veterinary reference laboratory involves the evaluation of a blood sample for the presence of von Willebrand's factor, a protein that assists in the adherence or 'stickiness' of platelets to sites of blood vessel injury. Von Willebrand's factor also helps to stabilise one of the coagulation factors (Factor VIII) in circulation so that this factor is not lost prematurely. Von Willebrand's disease is one of the most common hereditary bleeding disorders of dogs, affecting over 50 breeds and resulting in variable deficiencies of this factor.

"Von Willebrand's disease is one of the most common hereditary bleeding disorders of dogs..."

In addition to von Willebrand's factor testing, can the individual coagulation factors be evaluated?

Yes. Specific coagulation factor analysis is not commonly done, but it is available in Specialised veterinary or hematological laboratories. Specific coagulation factor analysis is typically used to evaluate the most common inherited deficiencies, namely von Willebrand's factor, factor VIII deficiency (hemophilia A) and factor IX deficiency (hemophilia B).

"Coagulation testing is not usually done in isolation."

Coagulation testing is not usually done in isolation. Typically, a CBC and a biochemical profile are first evaluated. Then a coagulation panel consisting of a combination of PT, PTT and possibly von Willebrand's factor testing is performed. The *combination of results* from these coagulation tests will indicate what part of the coagulation pathway is affected. Once these results have been assessed, the necessity for additional testing can be determined.

Antithrombin III is one of the most commonly measured inhibitors of the coagulation system. While measurement of this protein requires only a single blood sample, antithrombin III determination is performed only at Specialised laboratories. Antithrombin III measurement is usually undertaken if a pet is at risk of increased coagulability (increased chance of clot formation). Increased coagulation of blood may be a complicating factor of many diseases including cancer, severe inflammation, Cushing's disease, and kidney disease.

HYPERCOAGULABILITY IN ANIMALS

Hypercoagulability has several causes, but in essence it reflects a greater amount of procoagulants than anticoagulants in the blood. This means that the blood coagulates (clots) abnormally more than usual. Blood platelets are minute, disc shaped cell fragments in the blood that are responsible for clotting the blood. Too many active platelets, or too few active platelets, can result in severe health disorders. One of the possible causes of hypercoagulability is when there are too many active platelets in the blood. The end result of hypercoagulability is an episode of *thrombosis*, where clots will get trapped in arteries, veins, or in the heart, causing a loss of blood to the areas these arteries feed. Hypercoagulability is usually secondary to an underlying disease.

Symptoms and Types

A blood clot that is blocking arteries in the lungs will present as severe breathing difficulties that come on suddenly, rapid breathing, lack of energy, and possible fever. A blood clot blocking the aortic *artery* – the major artery from the heart to the body – will show as sudden weakness or paralysis, pain in the limbs, an absent or weak pulse in the arteries on the inside of the thigh, cold limbs, or blue-purple coloured nails

Causes

Hypercoagulability onset may be due to blood platelets that are stickier than normal, resulting in too much clotting of blood cells; deficiencies in antithrombin, a natural *anticoagulant* that prevents clotting in the arteries and veins; decreased removal of coagulation factors, that is, not enough procoagulants are being removed, resulting in an abundance of coagulating factors; or, defective fibrinolysis. Fibrinolysis is the process where fibrin, the protein end product of blood coagulation, is dissolved, resulting in the removal of small blood clots from the bloodstream.

Some of the other causes for this condition are:

- Protein-losing nephropathy: a condition resulting in a loss of protein from the intestines so that anti-clotting/anticoagulant proteins are lacking
- Immune-mediated *hemolytic anemia*: the abnormal breakdown of blood vessels
- Disseminated intravascular coagulopathy: a serious disease of the blood vessels, which is usually secondary to a life-threatening illness, or precipitated by a bacterial infection in the blood (*sepsis*)
- Inflammation of the *pancreas*
- Parasite infection: specifically with Dirofilaria, a *genus* of worms that includes the heartworm
- Under-functioning *thyroid gland*
- Cancer.

Diagnosis

Since hypercoagulability is usually the result of an underlying disease, your veterinarian will order tests for activated partial thromboplastin time (APTT) to measure how quickly your dog's blood is clotting, in addition to a blood chemical profile, a complete blood count, and a comprehensive *urinalysis*. Chest x-rays will help to visualise abnormalities in the lungs, and an abdominal ultrasound may be used to further examine the body for an aortic artery blockage.

An echocardiogram can also be used, for diagnosing blood clots in the heart, and for detecting high *blood pressure* in the lungs that may be present due to blood clots there. An examination using an injection of a *radiopaque* substance in order to view the blood vessels, called angiography, may also be necessary for confirming a blood clot. Another test, called nuclear *perfusion* scintigraphy, uses a radioactive tracer to illuminate the internal body, and is useful for non-invasively diagnosing blood clots in the lungs.

ANTICOAGULANT RODENTICIDE POISONING IN DOGS

The purpose of an *anticoagulant* is to prevent the coagulation (clotting) of blood. These agents are commonly used in rat and mouse poisons, and are one of the most common household poisons, accounting for a large number of accidental poisoning among dogs.

When ingested by an animal, anticoagulants block the synthesis of vitamin K, an essential component for normal blood clotting, which results in spontaneous and uncontrolled bleeding. Normally, dogs that have mild anticoagulant poisoning will not show signs of poisoning for several days, but as the poison begins to affect the system, the dog will become weak and pale due to blood loss.

The bleeding may be external; this may be displayed as a nose bleed, bloody vomit, or bleeding from the *rectum*. Dogs can also suffer from unseen internal bleeding; bleeding into the chest or abdomen, for example, is fatal if it not diagnosed in time.

Symptoms and Types

Here are some of the most common symptoms of anticoagulant poisoning:

- Weak, wobbly, unstable
- Nose bleeds
- Blood in vomit
- Blood in stools
- Bleeding from rectum
- Bruises and hematomas under the skin
- Hemorrhages (excessive bleeding) in the gums
- Ascites (swelling of the belly) due to accumulation of blood in the abdomen
- Difficulty in breathing due to blood in the lungs (this will make a rattling or crackling sound)

Causes

The main cause of anticoagulant poisoning is from the ingestion of rodent poison. If you suspect that your dog has come into contact with rat or mouse poison, and you are seeing some of the symptoms listed above, you will need to bring your dog to a veterinarian before your pet's health becomes critical. Keep in mind that outdoor dogs (or dogs that go outside frequently) are at risk of rodent poisoning.

It might be in a neighbour's yard, in a trash bag, or in an alleyway. Dogs that engage in chasing and killing rodents may also be susceptible to this type of poisoning. Even if you do not live in an area where rats or mice are a concern, rodent poison may be used for other common suburban pests like raccoons, opossums, or squirrels.

Some of the main anticoagulant chemicals that can be found in rodent poisons (or other household products) are:

- Warfarin
- Hydroxycoumadin
- Brodifacoum
- Bromadiolone
- Pindone
- Diphacinone
- Diphenadione
- Chlorohacinone

The first kind of anticoagulants are cumulative poisons. These poisons contain warfarin and hydroxycoumadin as main anticoagulants and they require multiple feedings that take several days to kill a rodent.

The second type of anticoagulant is deadlier, killing rodents in a single serving dose rather than over time. These deadlier anticoagulant poisons contain indanedione class products, like pindone, diphacinone, diphenadione and chlorohacinone, all of which are extremely toxic. Rodenticides that contain the ingredients bromadiolone and brodifacoum, for example, are 50 to 200 times more poisonous than the kind that contain warfarin and hydroxycoumadin.

Another cause of anticoagulant poisoning in dogs is the accidental ingestion of medication. Heparin, a common drug for treating blood clotting in humans, can have a toxic effect on animals. Often, dogs that have access to medications will eat what they have found, either because the drugs are within reach, or because the drug cabinet is kept unlocked.

Diagnosis

Your veterinarian will perform a thorough physical exam on your dog, taking into account the background history of symptoms and possible incidents that might have precipitated this condition. You will need to give a thorough history of your dog's health and recent activities. A complete blood profile will be conducted, including a chemical blood profile, a complete blood count, and a *urinalysis*. In addition, your doctor will check the time it takes the dog's blood to clot to determine to severity of the poisoning.

If you have a sample of the poison, you will need to take that with you to the veterinarian's office; bringing samples of the dog's vomit and/or stool may also be helpful.

Treatment

If your dog is suffering from spontaneous bleeding caused by anticoagulants, the treatment will involve administering fresh whole blood, or frozen plasma, in an amount determined by the rate and volume of the animal's blood loss. Vitamin K, which is necessary for normal blood clotting, will be used specifically as an *antidote*, and will be given by *subcutaneous* (under the skin) injections, with repeated doses as necessary — by injection or even orally — until the blood clotting time returns to normal. Do not induce vomiting unless you have been advised to do so by your veterinarian. Some poisons can cause more harm coming back through the *esophagus* than they did going down.

Living and Management

If your dog consumes a mild cumulative form of an anticoagulant, your dog may recover in a week, but if it was the lethal single dose anticoagulant, it may take up to a month.

Prevention

Anticoagulant poisoning can be prevented by keeping all poisons out of the reach of your pets. All other chemicals, drugs, and medications (especially blood thinning drugs) should also be kept out of your dog's reach — ideally inside of a cabinet.

ATRIOVENTRICULAR CONDUCTION

At physiological conditions an impulse is initiated by the sinus node and then passes successively through the atrial myocardium to the atrioventricular (AV) node, the His bundle, the right and the left bundle branches, and the Purkinje fibres in the ventricular myocardium. A basal heart rate is set by this intrinsic conduction system and is regulated by the autonomic (sympathetic and parasympathetic) innervation of the heart. The effects of NO on the heart rate are complex. Exogenous and endogenous NO seem to have different actions on the basal heart rate. In the isolated animal heart, NO appears to have a dose-dependent chronotropic effect on the sinus node. Sodium nitroprusside, a NO donor, has a biphasic chronotropic effect; it increases the beating rate of the guinea pig sinus node at a low concentration and decreases the heart rate at a high concentration. The action of NO on the sinus node is believed to result from the stimulation of the hyperpolarisation-activated inward current, I_f, in the atrial and sinus tissues. In the isolated rat heart, sodium nitroprusside (10 µmol/L) increases the average heart rate by 56 beats/min. Another NO donor, 3-morpholinosydonimine, had little effect on the beating rate of the isolated rat right atrium at low concentrations but reduced the rate at very high concentrations.

In the intact animal heart, the NO donors molsidomine and sodium nitroprusside have been found to elicit a linear increase in heart rate in anesthetised rabbit after cardiac autonomic denervation and beta-adrenergic blockade. In healthy subjects, sodium nitroprusside (2 µg/kg/min) causes some 12 per cent increase in heart rate in the absence of changes in brachial and aortic blood pressure. The effect of endogenous NO on the intrinsic heart rate is controversial. Manning *et al* showed that long term blockade of NO release with N^{ω}-nitro-L-arginine methyl ester (L-NAME) in dogs can cause hypertension and bradycardia. Heart rate variability, measured as a 24th standard deviation, was significantly diminished during chronic L-NAME administration. Hypertension, but not bradycardia, was reversed by L-arginine, a NO precursor. One may argue that the alteration in heart rate in the above study is difficult to interpret because of the presence of the baroreflex, which may itself influence the basal heart rate following L-NAME administration and the subsequent increase in blood pressure.

However, the changes in heart rate following endogenous NO suppression may be explained by the alteration of baroreflex sensitivity, and reversal of hypertension with L-NAME was not accompanied by recovery of bradycardia. Also, chronic NO synthase (NOS) inhibition does not change baroreflex sensitivity. Therefore, bradycardia during long term L-NAME treatment is likely the result of NO inhibition. In pentobarbital-anesthetised dogs, in which cardiac innervation and baroreflex were blocked by atropine and propranolol, infusion of a NOS inhibitor, N^G-monomethyl-L-arginine (L-NMMA), into the sinus artery had no appreciable effect on the basic heart rate. In pentobarbital anesthetised sheep, an increase in endogenous NO synthesis during high pressure intracoronary perfusion had no significant effect on heart rate but prolonged ventricular repolarisation without changing the heart rate.

NO does not appear to play a critical part in regulating AV nodal conduction in intact animals at physiological conditions. Inhibition of basal NO production following infusion of L-NMMA into dog AV nodal artery did not alter the A-H interval, a measurement of AV nodal conductivity. In anesthetised sheep, intravenous infusion of L-arginine did not cause appreciable changes in the PR interval on body surface electrocardiogram. However, NO seems to modulate the AV node's responses to autonomic stimulation. In isolated rabbit AV nodal cells, NO was found to mediate adenosine's suppressive effect on an isoprenaline-induced increase in the inward calcium current through a cGMP pathway. In intact dog heart, infusion of L-NMMA into the AV nodal artery attenuated the negative chronotropic and dromotropic responses of these tissues to vagal nerve stimulation. Intravenous administration of L-arginine reversed these responses towards the control values. In addition, NO inhibition with L-NMMA enhanced the effects of sympathetic stimulation or isoproterenol infusion on AV conduction, indicating that basal NO suppresses the stimulative effect of sympathetic nerves on the heart.

EFFECT OF NITRIC OXIDE ON VENTRICULAR REPOLARISATION

Under physiological conditions, basal NO does not appear to have a significant influence on ventricular repolarisation. In pentobarbital-anesthetised, open chest sheep, intravenous infusion of N^G-nitro-L-arginine did not change the epicardial activation-recovery interval, a well established measure of ventricular repolarisation in the *in vivo* heart, or the QT interval on body surface electrocardiogram, another indicator of ventricular repolarisation. In dogs, NO inhibition with pericardial application of L-NMMA exerted little change in the ventricular effective refractory period.

Although basal NO has little effect on ventricular repolarisation, increased NO synthesis or release prolongs it. In anesthetised sheep, high pressure intracoronary perfusion, which inevitably increases NO production, led to a small but significant prolongation in mean ventricular fibrillation intervals and activation-recovery intervals. NO also mediates the changes in ventricular repolarisation during sympathetic stimulation. In the intact dog heart, pericardial L-arginine administration and a subsequent increase in the systemic content of NO reduced the shortening of ventricular effective refractory periods during sympathetic stimulation. This effect was completely abolished by simultaneous administration of L-NMMA.

Nitric Oxide and Cardiac Arrhythmias

Acute myocardial ischemia or reperfusion of the ischemic myocardium results in fatal ventricular arrhythmias such as ventricular tachycardia or fibrillation. NO has two actions that may potentially prevent ventricular arrhythmias induced by myocardial ischemia or reperfusion. It dilates coronary arteries, which may reduce the extent of ischemia, and suppresses platelet adhesion and aggregation, which may prevent thrombosis within the coronary arteries and therefore the recurrence of ischemia. In addition, NO has some favourable effects on ventricular repolarisation, all of which may interrupt the formation of ventricular arrhythmias. However, recent reports on the potential antiarrhythmic effect of NO have been inconsistent and sometimes contradictory. Pabla and Curtis showed that NO behaves as an endogenous antifibrillatory factor in the isolated rat heart during reperfusion following sustained ischemia. L-NAME, which blocks NOS, increased the incidence of reperfusion-induced ventricular fibrillation from 5 per cent in the control condition to 35 per cent after 60 min of ischemia. The profibrillatory effect of L-NAME was prevented in hearts co-perfused with L-arginine. Furthermore, the proarrhythmic effect of L-NAME was prevented by pretreatment with a NO donor, sodium nitroprusside.

The above beneficial effects of NO on isolated rat heart have not been reproduced in the intact rat heart. In pentobarbital-anesthetised rats, sodium nitroprusside had no prophylactic effects on ventricular tachycardia, ventricular fibrillation or mortality resulting from 25 min myocardial ischemia. Another NO donor, 3-morpholinosydnonimine-N-ethylcarbamide, did not suppress the incidence or severity of ischemia-induced arrhythmias either. The authors of this study concluded that NO donors do not prevent arrhythmias induced by acute coronary artery occlusion or reperfusion in anesthetised rats. In a recent study, Sun and Wainwright showed that C87-3754, also a NO donor, caused a significant reduction in arterial blood pressure in anesthetised rats without reducing the incidence or severity of ventricular arrhythmias during a 30 min myocardial ischemia.

Recent studies of intact rat hearts have shown that NO may mediate the antiarrhythmic effect of certain chemical compounds. Resveratrol, an antioxidant in red wines, reduces the incidence and duration of ventricular tachycardia and fibrillation induced by ischemia and reperfusion. This antiarrhythmic effect is associated with an increase in plasma NO concentration in these animals. Suppression of NO synthesis with L-NAME has been found to abolish the antiarrhythmic effects of a newly discovered compound, honokiol, and its isomer, magnolol, in rats with 30 min coronary ligation.

Other evidence indicates that instead of being antiarrhythmic, NO may actually facilitate the occurrence of arrhythmias during myocardial ischemia in rat hearts. Naseem *et al* reported that NO inhibition with N^G-nitro-L-arginine, a NOS inhibitor, improves the left ventricular function and reduces the incidence of ventricular arrhythmias induced by ischemia or reperfusion. The duration of the sinus rhythm, or the arrhythmia-free period, in the N^G-nitro-L-arginine-treated group was also prolonged.

Ferdinandy *et al* showed that in the isolated rat heart, pretreatment with tetrakis-2-pyridylmethyl-ethylenediamine, a potent metal chelator, reduced basal cardiac NO content and prevented the accumulation of NO during ischemia-reperfusion. The incidence of ischemia- and reperfusion-induced ventricular fibrillation and ventricular tachycardia was significantly reduced following NO inhibition. In line with this study, Liu *et al* found that, in the in situ rat heart, myocardial ischemia and reperfusion-elicited arrhythmias were accompanied by a 90 per cent increase in the NO content in the ischemic tissues, along with a sixfold rise in inducible NOS activity in the same tissue.

The reports on the antiarrhythmic effect of NO in large animals are less controversial than those in small animal models such as rats. In anesthetised dogs, inhibition of basal NO resulted in a reduction in the antifibrillatory effect elicited by two 'preconditioning' occlusions of the left anterior descending coronary artery, implicating NO as an endogenous mediator of preconditioning against ventricular fibrillation. Also in anesthetised dogs, aminoguanidine, a NOS inhibitor, attenuated the cardiac protection against ischemic arrhythmias afforded by cardiac pacing. In chloralose-anesthetised pigs, intravenous administration of pirsidomine (1 mg/kg), a NO donor, reduced the incidence of ventricular ectopic beats following occlusion of the left anterior descending coronary artery. Although the incidence of ventricular fibrillation was unaffected by pirsidomine, the time to onset of this arrhythmia was significantly prolonged by this intervention.

In open chest dogs, pericardial application of L-arginine reduced the severity of ventricular arrhythmias induced by sympathetic stimulation on top of coronary occlusion. Pretreatment with the NO donors nicorandil and isosorbide-2-mononitrate reduced the incidence and severity of ventricular arrhythmias following occlusion of the left descending coronary artery in anesthetised dogs (28,29).

NO seems to mediate some drug-induced arrhythmias. NOS enzymes are essential to the control of NO synthesis. A constitutive form of NOS, eNOS, is expressed in myocytes and regulates the synthesis of NO in these cells. A higher rate of ouabain-induced ventricular arrhythmias was found in ventricular myocytes isolated from mice lacking a functional eNOS gene.

Application of a NO donor, *S*-nitrosoacetylcysteine, diminished the drug-induced arrhythmias in these preparations. Furthermore, in isolated guinea pig heart, pretreatment with sodium nitroprusside and L-arginine reduced the incidence of digoxin-induced ventricular arrhythmia. Sodium nitroprusside, but not L-arginine, was also found to suppress digoxin-induced ventricular arrhythmias in intact guinea pig heart.

Potential Mechanisms Changes in Cardiac Electrophysiology

So far the electrophysiological properties of NO have mostly been studied using NO donors or NOS inhibitors. The actions of NO in the myocardium are extremely complex, with myocytes or cardiac neurons being the potential sites of action. It is almost impossible, at this stage, to show the exact mechanisms of action in the myocardium. NO has been shown to mediate intracellular production of cyclic GMP through guanylyl cyclase, which may account for many of the observed electrophysiological actions of NO. NO has also been shown to inhibit slow inward calcium currents through cyclic GMP-dependent protein kinase. How these molecular effects may contribute to the observed electrophysiological effects of NO is still unclear.

Because NO dilates coronary arteries and suppresses platelet aggregation, a NO-mediated decrease in ventricular arrhythmias following myocardial ischemia may be due to improvement in myocardial perfusion in addition to the favourable changes in ventricular repolarisation. Several studies have shown that L-arginine and NO donors diminish ischemia-reperfusion injury and improve postischemic mechanical function, whereas NOS inhibitors are associated with a poor recovery from the ischemia-reperfusion damage.

CIRCULATORY SYSTEM OF ANIMALS

The circulatory system is the continuous system of tubes through which the blood is pumped around the body. It supplies the tissues with their requirements and removes waste products. In mammals and birds the blood circulates through two separate systems - the first from the heart to the lungs and back to the heart again (the pulmonary circulation) and the second from the heart to the head and body and back again.

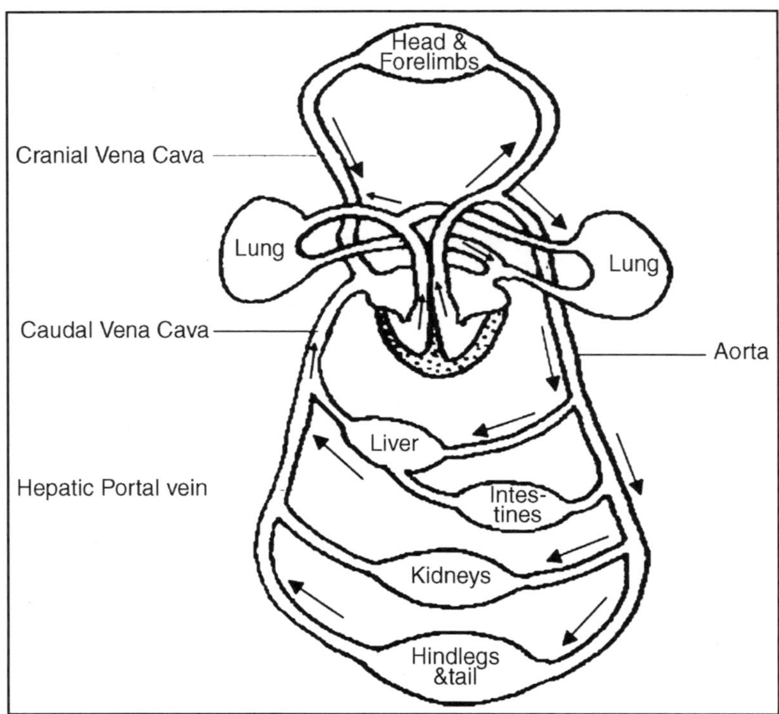

Fig. The Mammalian Circulatory System

The tubes through which the blood flows are the arteries, capillaries and veins. The heart pumps blood into arteries that carry it away from the heart. The arteries divide into very thin vessels called capillaries that form a network between the cells of the body. The capillaries then join up again to make veins that return the blood to the heart.

Arteries

Arteries carry blood away from the heart. They have thick elastic walls that stretch and can withstand the surges of high pressure blood caused by the heartbeat. The arteries divide into smaller vessels called arterioles.

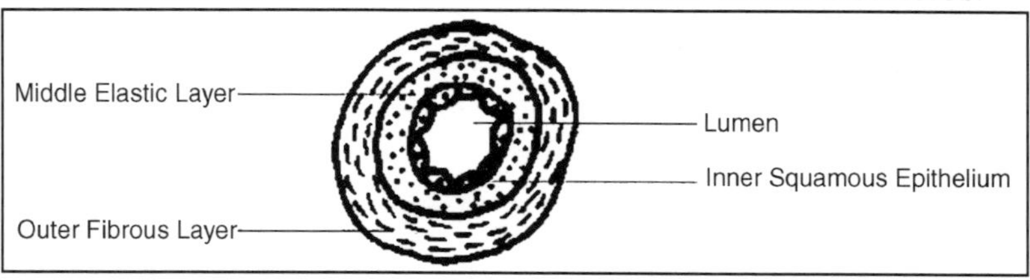

Fig. Cross Section of an Artery

The hole down the centre of the artery is called the lumen. There are three layers of tissue in the walls of an artery. It is lined with squamous epithelial cells. The middle layer is the thickest layer. It made of elastic fibres and smooth muscle to make it stretchy. The outer fibrous layer protects the artery. The pulse is only felt in arteries.

The Pulse

The pulse is the spurt of high pressure blood that passes along the aorta and arteries when the left ventricle contracts. As the pulse of blood passes along an artery the elastic walls stretch.

When the pulse has passed the walls contract and this helps push the blood along. The pulse is easily felt at certain places where an artery passes near the surface of the body. It is strongest near the heart and becomes weaker as it travels away from the heart. The pulse disappears altogether in the capillaries.

Capillaries

Arterioles divide repeatedly to form a network penetrating between the cells of all tissues of the body. These small vessels are called capillaries. The walls are only one cell thick and some capillaries are so narrow that red blood cells have to fold up to pass through them.

Capillaries form networks in tissues called capillary beds. The capillary networks in capillary beds are so dense that no living cell is far from its supply of oxygen and food.

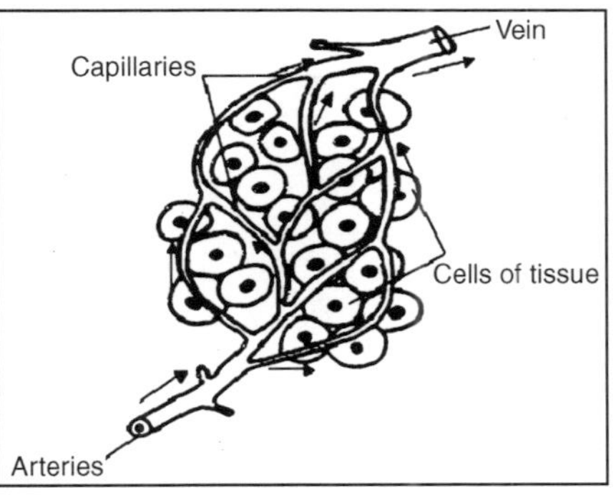

Fig. A Capillary Bed

The Formation of Tissue Fluid and Lymph

The thin walls of capillaries allow water, some white blood cells and many dissolved substances to diffuse through them. These form a clear fluid called tissue fluid (or extracellular fluid orinterstitial fluid) that surrounds the cells of the tissues. The tissue fluid allows oxygen and nutrients to pass from the blood to the cells and carbon dioxide and other waste products to be removed from the tissues.

Some tissue fluid finds its way back into the capillaries and some of it flows into the blind-ended lymphatic vessels that form a network in the tissues. Once the tissue fluid has entered the lymphatics it is called lymph although its composition remains the same. The lymph vessels have walls that are even thinner than the capillaries.

This means that molecules and particles that are larger than those that can pass into the blood stream *e.g.* cancer cells and bacteria can enter the lymphatic system. These are then filtered out as the lymph passes through lymph nodes.

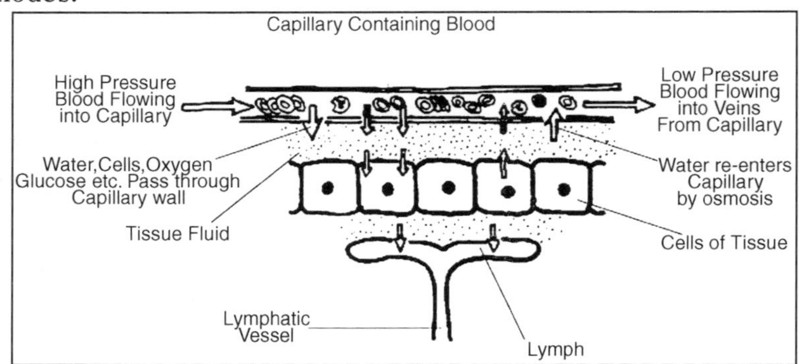

Fig. The Formation of Tissue Fluid and Lymph from Blood

Veins

Capillaries unite to form larger vessels called venules that join to form veins. Veins return blood to the heart and since blood that flows in veins has already passed through the fine capillaries, it flows slowly with no pulse and at low pressure.

For this reason veins have thinner walls than arteries although there have the same three layers in them as arteries. As there is no pulse in veins, the blood is squeezed along them by the contraction of the skeletal muscles that lay alongside them. Veins also have valves in them that prevent blood flowing backwards.

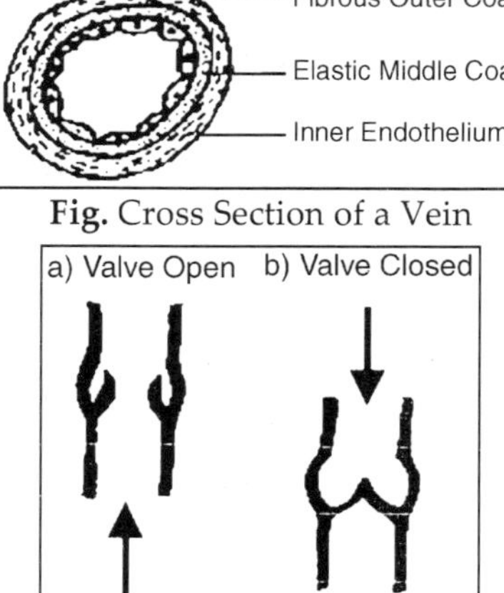

Fig. Cross Section of a Vein

Fig. a) and b) Valves in a Vein

REGULATION OF BLOOD FLOW

The flow of blood along arteries, arterioles and capillaries is not constant but can be controlled depending upon the requirements of the body. For example more blood is directed to the skeletal muscles, brain or digestive system when they are active. Regulation of the blood flow to the arterioles of the skin is also important in controlling body temperature. The size of the vessels is adjusted by the contraction or relaxation of smooth muscle fibres in their walls.

Oedema and Fluid Loss

Oedema is the swelling of the tissues due to the accumulation of tissue fluid. This may occur because the tissue fluid is prevented from returning to the bloodstream and accumulates in the tissues. This may be caused by physical inactivity (*e.g.* long car or plane trips in humans) or because of imbalances in the proteins in the blood. This is what causes the "pot-belly" of the malnourished child or worm-infested puppy.

Loss of body fluid can be caused not only by drinking insufficient liquid but also through diarrhea and vomiting or sudden loss of blood due to haemorrhage. The effect is to reduce the volume of the blood which decreases the blood pressure. This could be dangerous because the supply of adequate blood to the brain depends upon maintaining the blood pressure at a constant level. To compensate for the loss of fluid various mechanisms come into play. First of all the blood vessels contract in order to try and maintain the pressure. Then, since the loss of fluid tends to make the blood more concentrated and increases its osmotic pressure, fluid is drawn into the blood from the tissues by osmosis.

The Spleen of Stomach

The spleen is situated near the stomach. It has a rich blood supply and acts as a reservoir of red blood cells. When there is a sudden loss of blood, as happens when a haemorrhage occurs, the spleen contracts to release large numbers of red blood cells into the circulation. The spleen also destroys old red cells and makes new lymphocytes but it is not an essential organ because its removal in adult life seems to cause few problems. In the foetus, the spleen makes both red and white cells.

Important Blood Vessels of the Systemic (Body) Circulation

Blood is pumped out into the body via the main artery, the aorta. This takes the blood to the head, the limbs and all the body organs. After passing through a network of fine capillaries, the blood is returned to the heart in the largest vein, the vena cava. Arteries and veins to and from many organs often run alongside each other and have the same name *e.g.* the renal artery and vein serve the kidney, the femoral artery and vein serve the hind limbs and the subclavian artery and vein serve the forelimbs. However, blood to the head passes along the carotid artery and returns to the cranial vena cava via the jugular vein.

One variation on this arrangement is found in the blood vessels that serve the digestive tract.

A variety of arteries take blood from the aorta to the intestines but blood from the intestines is carried by the hepatic portal vein to the liver where the digested food can be processed. This vessel is unlike others in that it transports blood from one organ to another rather than to or from the heart like arteries or veins.

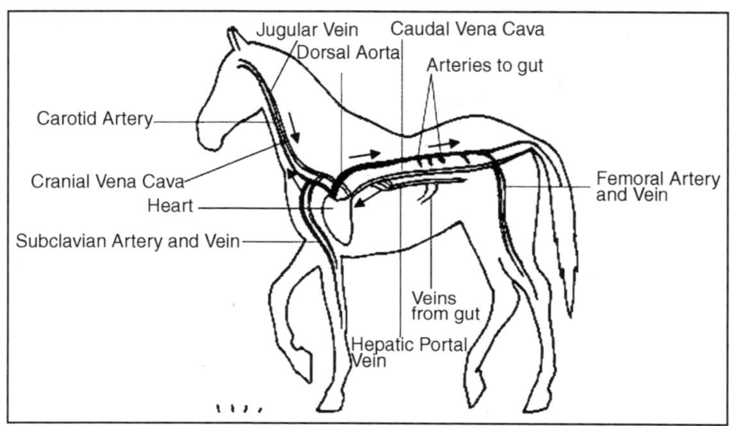

Fig. The Main Arteries and Veins of the Horse

Blood Pressure

The blood pressure is the pressure of the blood against the walls of the main arteries. The pressure is highest as the pulse produced by the contraction of the left ventricle passes along the artery.

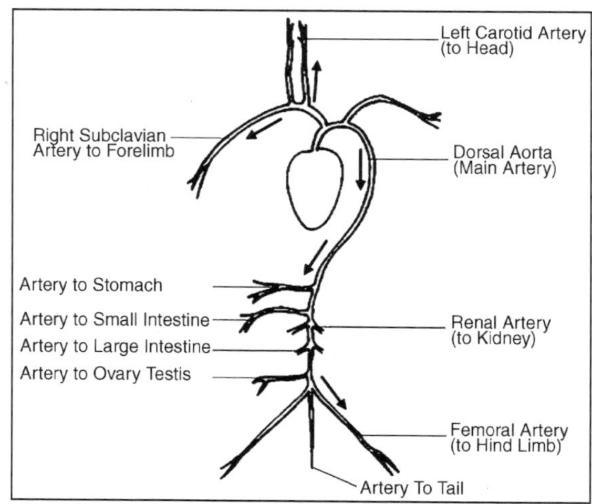

Fig. The Main Arteries of the Body

This is known as the systolic pressure. Pressure is much lower between pulses. This is known as the diastolic pressure. Blood pressure is measured in millimetres of mercury. A blood pressure that is higher than expected is known as hypertension while a pressure lower than expected is known as hypotension.

MODELLING ANIMAL CARDIAC ELECTROPHYSIOLOGY

Mathematical models can also be used to replicate the results of experiments carried out in animals, as well as investigating the mechanisms underlying heart function. The models could even be used to investigate conditions that would be impossible or prohibited using animals. The action potential wave form (AP) is represented by a graph and describes the changes in electrical potential of a cell.

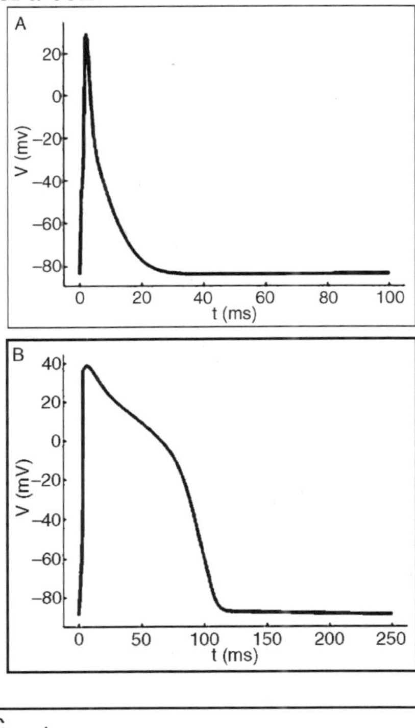

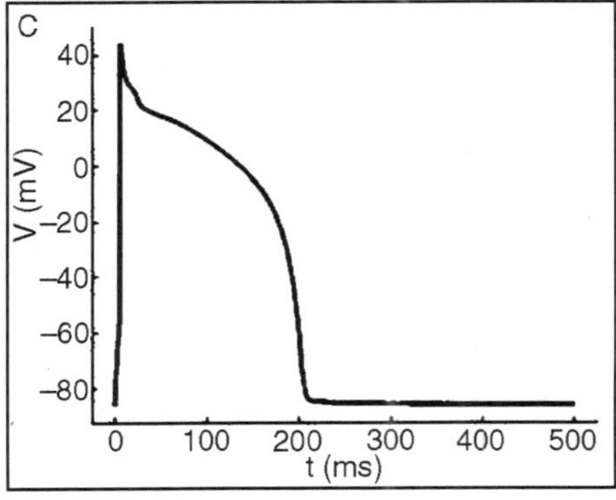

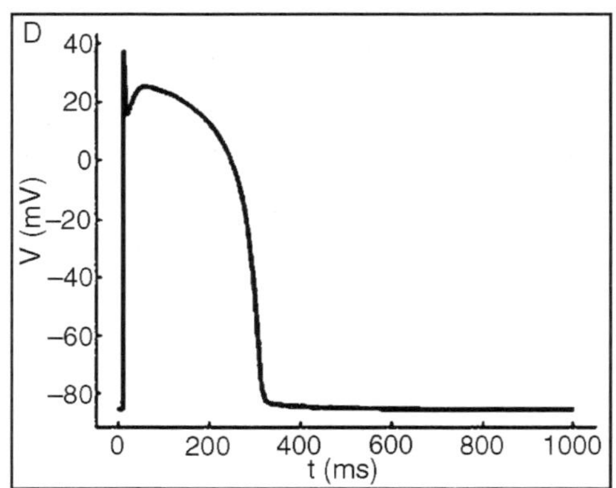

Fig. Electrical Waveform in Single Muscle Cells During One Heart Beat for Various Species: (A) Mouse, (B) Guinea Pig, (C) Rabbit, and (D) Human. Note the Different Shapes and Durations of the Wave Forms (Time in Milliseconds).

The action potential serves two purposes: it triggers the contraction of a cell; and it interacts with the electrical potential of neighbouring cells to coordinate their contractions.

APs differ between species. The APs of four different species: mouse, guinea pig, rabbit, and human (Panels A to D, respectively). The APs of the four species are very different from each other, which makes it difficult to draw conclusions from experiments carried out in one species when the results are compared with another. The different amounts of specific ion channels expressed in each species account for these apparent differences in the waveforms.

Mathematical modelling can be used to compare these species-specific experimental results.

For example, "knocking out" (eliminating) the sodium-calcium exchanger gene NCX, which prevents any active protein from being produced, did not have substantial effects in mice, but mathematical models could show that it would be much more dramatic and possibly life threatening if NCX did not function in larger animals that have a longer AP and a substantial plateau phase.

APs that maintain positive membrane potentials for longer, producing a wave form that is a plateau. In mice and other small animals that have short APs, removing NCX does not change the AP waveform because other ion currents can take the place of those channelled through NCX. In larger animals, because NCX contributes substantially to maintaining the plateau phase, eliminating the protein would change the AP waveform dramatically by changing the balance of ions and ion currents within the heart.

Computer modelling can be used to investigate the effect of calcium overload on heart cells that occurs when the heart tissue is starved of oxygen (ischaemia; 4). For example, experiments and mathematical models have both demonstrated that blocking the action of the NCX protein using pharmaceutical compounds results in the accumulation of intracellular calcium causing abnormal contractions of heart muscle.

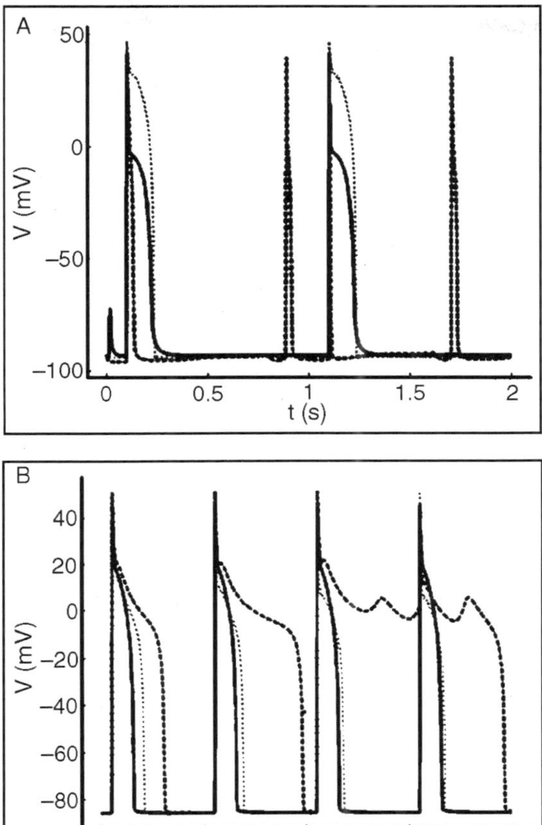

Fig. Blocking the Sodium-calcium Exchanger Protein (NCX) Might not be Beneficial in Ischemia When the Main Risk is Intracellular Calcium Build Up that Disturbs the Electrical Waveform Pattern (Compare Solid and Dashed Lines in Graph A). Blocking the Action of NCX Might help to Normalise the Electrical Waveform in Patients with "Long QT Syndrome" (Graph B). Solid Line Denotes the Control (Normal) Case, Dashed Line the Disease State, and the Dotted Line Indicates the Effect of Blocking the Action of NCX.

The same mathematical model was used to show that another type of arrhythmia is caused when repolarisation fails to occur in people with genetic mutations or in people who are taking specific pharmaceutical drugs.

Many drug compounds can do this at high concentrations, creating a safety concern for the pharmaceutical industries and health authorities. The mathematical model showed that blocking the action of NCX by 50-80 per cent in these circumstances could be beneficial and could restore the AP length to normal in these patients. The model showed that this was mainly due to the reduced inward flow of calcium ions as a result of the blocked action of NCX, which resulted in a shorter plateau phase allowing the calcium ion levels and the resulting AP to return to normal.

COMPUTER MODELS OF THE WHOLE VENTRICLE

Mathematical modelling can be used to simulate a healthy heart and compare it with a model of an ischaemic one. In addition, the simulations can be used to study the effect of drugs, defibrillation and cardiac pacemakers, and to optimise drug treatments. Models exist that represent the whole ventricle of the heart. In order to do this millions of cells have to be included in the simulations and high performance computers are necessary to model just 3 or 4 heart beats. It currently still takes days to do this, however improvements to computational speed will enable more complex and detailed in silico investigations of the heart.

Using Computer Models for Hypothesis Testing

Cardiac modelling can also be used to test current hypotheses. For example, researchers used mathematical modelling to determine whether or not the changes in expression levels of four ion channels observed in experiments were sufficient to explain the changes in AP waveform and intracellular calcium dynamics observed.

By adjusting the ion currents in the mathematical model to mimic the experimental situation, they observed a change in AP waveform equivalent to the experimental results seen. If these adjustments to the model had not been sufficient to confirm the experimental results, a parameter analysis of the model would have been conducted, which might have revealed other changes in gene expression levels; this information would then have been used to influence the design of further experiments to validate the model prediction.

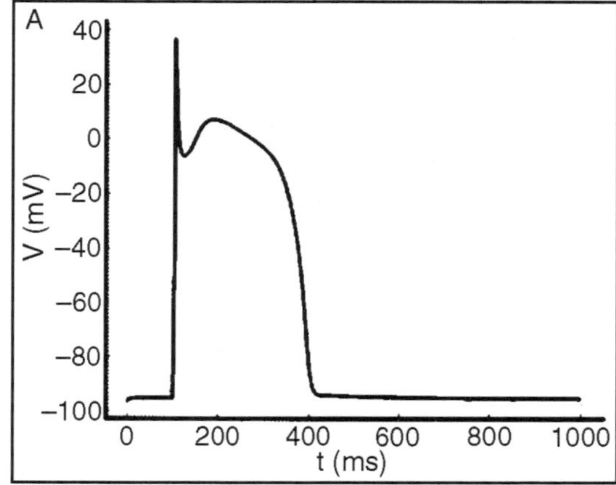

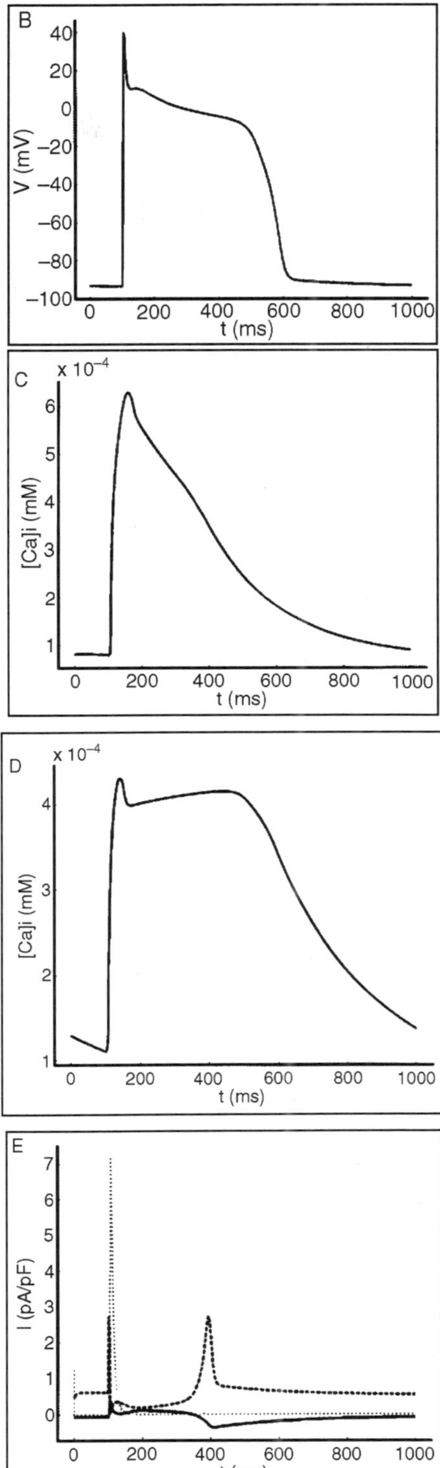
B
V (mV)
40
20
0
−20
−40
−60
−80
−100
0
200
400
600
800
1000
t (ms)
C
x 10^−4
[Ca]i (mM)
6
5
4
3
2
1
0
200
400
600
800
1000
t (ms)
D
x 10^−4
[Ca]i (mM)
4
3
2
1
0
200
400
600
800
1000
t (ms)
E
I (pA/pF)
7
6
5
4
3
2
1
0
0
200
400
600
800
1000
t (ms)

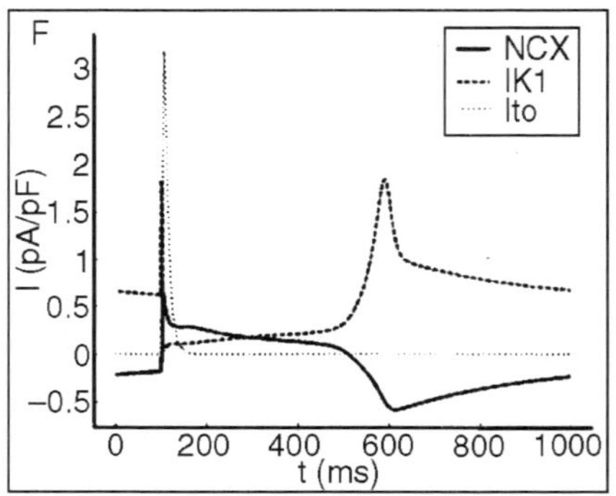

Fig. The Action Potential Waveforms of Normal Cells and Heart Failure Cells are Shown in Graphs A and B, Respectively; Intracellular Calcium Dynamics in these Cells is Shown in Graphs C and D. The Underlying ion Currents Responsible for these Changes are Shown in Panels E and F. Heart Failure (All Graphs on the Right) is Accompanied by Changes in the Amounts of Specific ion Channels in the Heart Muscle Cells. Mathematical Modelling Shows that Expression Level Changes of ion Channel Proteins are Sufficient to Explain the Differences in Electrical Waveform Compared with a Normal Heart (Left Panels).

Cardiac Modelling and Drug Development

Cardiac disease and cancer are two of the main causes of death in the western world; consequently substantial effort is put into researching new drug compounds that could reduce mortality from these diseases. However, more than half of the new compounds that might be beneficial for treating cancer, diabetes, etc. have adverse side effects on the heart.

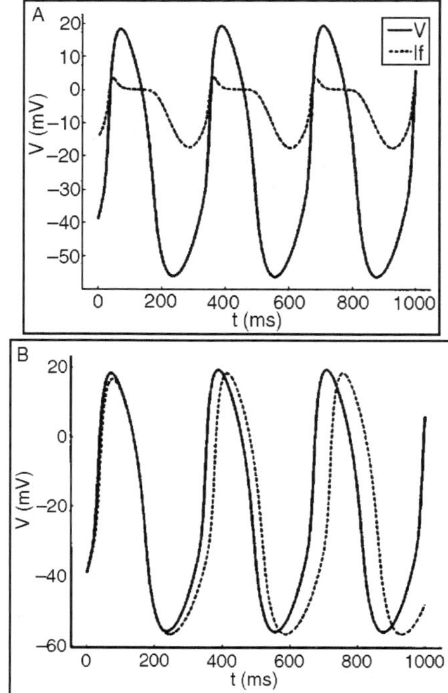

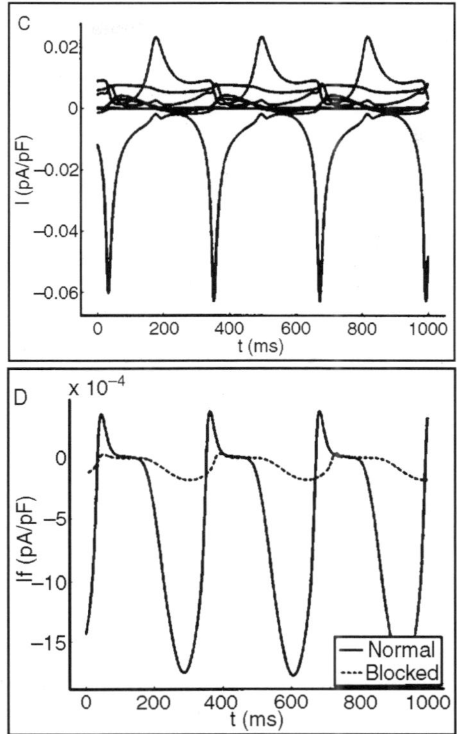

Fig. In a Mathematical Model of a Pacemaker Cell the Hyperpolarising Current (If) was Shown to be the Main Current Active During the Depolarising Phase that Determines Heart Rate. Graph A Shows the Membrane Potential and the Current If, Whereas Graph C shows the Main Underlying Currents with If. Using Ivabradine to Block this Current (Shown in Graph D) Sows Down the Heart Rate (Solid vs. Dashed Line Graph B).

Electrophysiological models of heart function can be used to improve the safety and efficacy of drugs. For example, Ivabradine, a drug used to treat angina, was developed using computer modelling and was approved by the European Medicines Agency in 2005. Researchers studying rabbit cardiac pacemaker cells (cells that control heart rate by creating the rhythmical electrical impulses), discovered that these cells produced another, previously unknown electrical impulse current.

When this current was programmed into a newly developed mathematical model the investigators found that it was the main current bringing sodium and potassium ions into the cell during the phase immediately following the initiation of a heart beat.

As this phase is important for determining the heart rate a change in this ion current translates into a change in heart rate. Because this ion channel protein was thought to be expressed almost exclusively in pacemaker cells, it represented an excellent target for drug compounds that decrease heart rate, and could therefore be useful for developing treatments for angina.

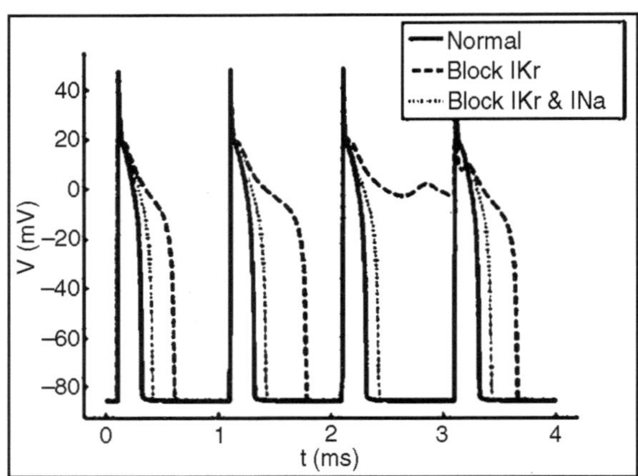

Fig. Mathematical Modelling has been used to Understand the Mode of Action of Ranolazine. Even Though the Drug Blocks One of the Important Repolarising Currents (IKr) - Which on its own could Lead to Sudden Cardiac Death (Dashed Line) due to its Action on other Currents (Especially the Persistent Sodium Current) it Works as an Anti-arrhythmic to Normalise the Currents (Dotted Line).

Mathematical modelling was also used in developing another drug used to treat angina, Ranolazine. This compound prolongs action potential, by blocking one of the important potassium ion repolarising currents (IKr) however, due to its activity on multiple ion channels, in particular the persistent sodium current, its overall effect is actually anti-arrhythmic.

Mathematical modelling could therefore be used to better understand the mode of action of Ranolazine and to gather the necessary data to convince the scientific community and the regulatory agencies that its effect of prolonging the AP does not necessarily mean that the compound will initiate arrhythmias.

6

Important Diseases of Animals

There are two types of diseases mainly seen in animals. a) contagious b) infectious.

- *Contagious:* As the name its elf suggests, contagious diseases are those which spread by contact/nearness of the animal.
- *Infectious:* These are diseases which are caused due to infective agent. (Virus or Bacteria or Fungus).All contagious diseases are infectious but all infectious deseases are not contagious

Diseases can also be classified as Viral Bacterial,Fungal,Rickettsial,Or Protozoon,Viral Deseases Important ones.Rinderpest,Foot and mouth,Ephimoral Disease, Blue tongue, Infectious bovine rhinotracheitis Malignant Catarrhal fever, Rabies,

- *Rinderpest disease:* Causes diarrhoea, high fever and death. Effective vaccine control has controlled this disease in animals since 1998. There are no outbreaks recorded since them.
- *Foot and Mouth Disease:* It is a viral disease. It has four main strains observed in India.(A,O,C,Asia I). In the symptoms animal goes lame. There is profuse salivation, animal goes lame. This disease has severe economic impact, on animal production (milk and Traction/Transport) This disease does not result in mortality (except few cases in calves). Disease can be prevented by undertaking regular vaccination. There are two types of vaccines. Aluminium hydroxide Gel and oil adjuvant vaccine. Both the vaccines are effective and useful.1st dose of vaccine is given at 2-3 month age. There after vaccine is given every 3 months as per the advice of the Veterinarian.

EPHIMERAL DISEASE

This viral disease is transmitted by insects. Animal shows high fever (104-105 F) and pain in muscles and joints. Animal is unable to walk, eat and ruminate. In India there is no vaccine against this disease. Symptomatic treatment such as non- steroidal anti inflammatory drugs, soda alycylus,, potassium iodide is given/venously. The course of the disease is for 4 to 7 days. Animal recovers of its own. Good nursing care is essential.

- *I.B.R INFECTIOUS BOVINE RHINOTRACHEITIS:* It is a new emerging disease causing abortion, repeat breeding and infertility. It is estimate that nearly 30 per cent of cross breads are affected. It shows various clinical symptoms. Respiratory disease (I B R) Genital disorders (IPV)infectious pustular vulvo-vaginitis in heifers/cows/ buffaloes and infectious pustular balanoposthitis (IPB) in bulls. Infections of eye, nervous symptoms, etc. In India it was first recorded in 1976. Diagnosis is done by serological examination. Prevention is done by using oil adjuvant vaccine.

Malignant Catarrhal Fever (MCF)

All ages of animals are suseptible. There is Eye and Nose discharges, high fever, gastroenteritis,dyspnoea. Muzzle is red. There is central nervous system involvement.

The disease is suspected to be transmitted by insects. Recovered animals may carry virus for some months. In per acute cases animal dies within one to three days.

In most common form there is drop in the milk. Acute gastro enteritis, swollen nasal mucous memberance high fever (104 to 105F),odema of eyelids, muzzle is red. There is foetid diarrhoea. There may be severe nervous dipression.

Necrosis of skin over udder and teats may also be seen. Urine may be red because of free lesion.In bladder. Disease runs a course of 3 to 7 days. There is no effective treatment. Oxytetracycline intravenously, vit A, and good nursing can be carried out. Quarantine and slaughter are suggested for control.

- RABIES: This viral disease is caused due to bite of an infected animal such as Dog,Cat,Fox,Bat or other domestic animal with a penetrating bitting wounds. Virus travels through nerves. Effective prevention is done by tissue culture vaccine in those areas where there is nearness of forest. There is no treatment.

BACTERIAL DISEASES

Amongst important diseases following one need prevention and effective control. Anthrax,black quarters, haemorrhagic septicaemia,

Brucellosis,isolation of animal and strict hygenic measures are needed to control Tuberculosis and johne's disease.

For remaining diseases antimicrobials can be tried:

- *Anthrax:* This disease is seen on very less number of occasions now a days after the invention of antibiotics esp. Penicillin. It is caused due to Bacillus Anthracis. Disease causes sudden death in animals without apparent symptoms. The carcase gets bloated. In contact animals can be given higher doses of Penicillin. Disease can be controlled by using spared vaccines as per Veterinarians advice in endemic areas. Where there is no disease recorded in last three years, clean vaccination is not advised by many State Govt authorities.

- *Black Quarters:* Disease is caused by claustridium charivoei, an anerobic organisms young healthy animals are affected. Infection is mainly through small wounds. There is swelling on the scapular muscles or on quarters, which on pressing gives crepitating sound. There is high fever and animal goes lame. Immediately use of antibiotics locally and parentrally is effective. Prevention can be done by carrying vaccination in endemic areas before mansoon.

HAEMORRHAGIC SEPTICAEMIA

This Disease is caused by organisms known as Pasteurella multocida. The infective agent enters through respiratory tract by inhalation or through feed/ water.

There is upper respiratory tract infection involving large bronchi and lunge. There is high fever,watery discharge from nostrils. Animal is off feed, there is odema under the neck.

There is respiratory distress. Animal dies if not treated in time. Mostly Buffalo species are affected. Higher antibiotics Tetracyclines,Sulpha Trimethoprim,or Enrofloxacin may be tried in early phases. No oral medication be done. Disease can be prevented by undertaking regular vaccination (GEL TYPE VACCINE) twice a day (Before mansoon and in october).

- *Brucellosis:* This disease is caused by Brucella abortus. It is a disease of matured animals. It causes great economic loss by causing abortions in pregnant animals. In bulls it causes orchitis. The causative agent enters into body mainly through food/water and through wound even through conjunctival mucous memberance. Organisms locate in gravid uterus and causes inflammation of cotyledons resulting into abartions. It has zoonotic importance.There is no effective treatment. Diagnosis is done by serological examination. Prevention is done by undertaking calf hood vaccination (in females) males are not vaccinatec. Serological examination of all animals isolation of affected animals and calf hood vaccination is advised.
- *Tuberculosis:* It is caused by Mycobacterium tuberculosis. Infeetion is mostly oral, either through feed or water. The disease has chronic course. Animal becomes weak day by day. In the udder form,milk is positive for organisms. Diagnosis is by all allergic tests carried out on skim. Positive animals are segregated. There is no treatment. It has zoonotic improtance.
- *John's Disease:* It is caused due to Mycobacterium para-tuber culosis. It has chronic couse. Infection is through oral route. Animal goes on Wasting. There is chronic diarrhoea. In pregnant animal disease

appears to be suppressed, how ever on calving the symptoms start again. Diagnosis is done on the basis of allergic tests. There is no treatment. Test, and segregate positive animals.

PROTOXOON DISEASE

These disease are carried by Ticks or bitling insects. The important disease are Theilariasis, Babesiosis, Anaplasmasis, Trypanasomiasisn.

Theilariasis: This is transmitted by Tciks. In India we have Theilaria annulata and Th.mutans. Th.annulata is pathogenic while Th.mutans is non- pathogenic. As the infective stage invades red blood cells, there is weakness,anaemia,and drop in production. The disease runs in Acute, sub-acuteand chronic form. There is initially high fever, local lymphriode swelling. Animal is dull. Diagnosis is done on blood smear examination. Treatment includes Butalex (Buparvaquone) 1.5mg/kg body wt 1/musular. Tetracyclines higher doses are also tried. Supportive treatment such as B.Complex,Iron,Dextrose are given. Blood transfusion/PCV transfusion is also effective.

Babesiosis: This is tick bowrne disease. We have Babesia bovis, B.bigermina. Animal shows high fever (uptoloyF). There is coffee coloured urine. Animal may die if not treated. Some animals, because of destruction of red blood cells may show jaundice.

Diagnosis is done on the examination of blood smears. For the treatment berenil (dimenazen) @ 7 mgs/kg sub cut is given. imidocarb 9imidocarb dipropionate) @1-2 mg per kg 1/muscular is also given.

Anaplasmosis: This is insect bourne disease. It has two types. anaplasma centrale and anaplasma marginale. There is high fever (104-105F). There is red coloured urine. Animal may show jaundice. Animal becomes anaernic. Diagnosis is done on blood smear examination. In the treatment higher doses of antibiotics.tetracyclines @ 10 mgs/kg.body wt 1/muscular is effective.

Trypanoso Miasis: This disease is transmitted by bitting insects from one to another animal. Infected animal shows high fever, there is reduction in production. There is odema over jawl on the brisket. In some animals there is central nervous system involvement. Animal moves in circle, it may dash against wall. Diagnosis is done on blood smear examination. Treatment can be based on one of the following. Berenil (dimenazen) @7 mg/kg s/c Antricide Prosalt 5 mg/kg wt s/c, Antricide sulphate @ 5 mg/kg s/c, samorin (isometamedium) 0.5 to 2 mg per kg s/c cymelarsan 0.3 to 0.6 mg/kg s/c control of insects by spraying acaricides is important.

PARASITIC DISEASES

Dairy animals suffer from different types of worms. They are Round Worms, Tape Worms, Flukes and Coccidiosis. Young animals suffer more as compared to adult animals. General symptoms include weakness,drop in production and reduced immunity. Intensity of symptoms depends upon worm load and its type.

Infection due to Liverfluke/Coccidiosis causes more damage when present. Diagnosis is done on the examination of stools. There are specific effective drugs for the treatment.

In broad spectrum anthelmintics. albendazol,oxfendazol,invermectin are suggested. They are effective. They need repeatation after three weeks. In case of coccidiosis sulpha drugs can be given. There are ectoparasitic diseases which infect skin.

They are due to lice/mites. They cause skin lesions. Loss of hair, irritation, suratching are the symptoms. Diagnosis of type of mite can be carried out by examination of skin scrapings. Specific drugs are given for specific infestation.

METABOLIC DISEASES

Dairy animals, due to their high yielding nature are likely to suffer from metabolic diseases. These are milk fever,ketosis,post parturient haemoglobin urea and hypomagneshimia.

- *Milk Fever:* There is no fever in this disease. This disease is seen mostly during the period of one week before calving to a week after calving. There is acute deficiency of circulating blood calcium. This results in depression, drop in milk yield, and recumbancy. Diagnosis is done on the basis of typical clinical symptoms and history. Treatment of calcium borogluconate 25 per centsolution 200 to 400 ml Intravenously gives quick results. Injection of ivt.D3 is also useful. Prevention can be carried out by anionic mixtures before calving for calcium absorption. They are Ammonium chloride @150 gms/day for 10 days before calving.
- *Ketosis:* In high yielding animals deficiency of stored energy results in utilising body fact. While energy is available in this process, certain other products such as Ketone bodies (keto acids) are formed. The disease usually occurs during the peak yield phase. Animal passes little urine,it has sweet odur to its breath,it drops in production and becomes weak. Diagnosis is done by examination of Urine,for ketone bodies. Now a days paper strips are available. In the treatment Dextrose 20-50 per cent 500 to 300 ml i/venously along with Gluco-Corticoids 1/muscularly 3-5 ml are effective. Use of propylene glycol (300-500 gms) orally once/twice a day is effective. Proper feeding in advance pregnancy for storage of energy prevents disease incidance.
- *Post Parturient Haemoglobinurea:* This disease occurs after calving due to acute deficiency of circulating phosphorus. There is lipid peroxidation of Red blood cell, resulting in its breaking and causing haemoglobinurea. Treatment consists of giving sodium acid phosphate 20 per cent 300-500 ml 1/venously along with oral dosing @100 gms/

day. Injection of Botropase, Paramind Benzoic Acid are also effective.

- Hypomagnesemia: Acute deficiency of magnesium, results in high yielding animals due to excessive intake of potassium (green tender leaves feeding during mansoon) Animal becomes restless, it collapses, it has froth at mouth, heart beats are loud. Immediate treatment by using 10 per cent Magnesium sulphate 100-200 ml I/venously will save the animal.
- *Mastitis:* It is the inflammation of the Udder. The causative agent may be Virus, Bacteria,Fungus or Algae. Non- infective cause is deficiency of circulating cirtrate. Causative agent enters through teat sphincter (opening) mostly after milking (by hand or Machine) When sphincter muscles are yet to close tightly. (45 minutes to 1 ½ hour after milking). Once entered, it has good media to multiply inside the udder. In per acute form animal may die. In Acute case there is high fever, udder is swollen and painful. There is watery milk. In sub acute cases there is little swelting, no fever but milk is not normal. In chronic cases, there is little abnormality in udder how-ever milk quantity as well as quality is depressed. The economic losses due to sub-clinical mastitis are more than clinical mastitis. In this form of disease there is no outward abnormality in the udder. Milk is normal but reduced. Per acute, acute,sub acute form of disease depends upon pathogenicity of organisms their number and defence mechanism status of the udder. As the causative agent enters through teat opening, minor cracks on teat or udder. It is therefore necessary to strictly follow practice of clean milk production. This includes washing of udder, drying (by cloth) of udder, before milking is necessary. Milking by full hand (not by following thumb) is also necessary. Milking when done by hand should be done quickly-completly and comfortably(comforts to cow). After milking all the teats should be dipped in new, non- toxic, non-irritant disinfectant such as INSTACLEAN (Benzalkonium chloride +0.1 per cent1-6dihexane) is useful it has long residual effect and works even in the presence of organic matter. If this practice is followed, incidance of sub-clinical mastitis can be controlled. In advanced pregnant animals, at the time of drying, all the teats should be infused with antibiotics. This will eliminate chronic mastitis.In the treatment following principles be observed.
- Palpate the udder know its feel (hotness,hardness)
- Record animals body temperature.
- Remove milk from affected milk as frequently as possible.
- If udder is hot and painful to touch,foment by ice/or cold water. Give antiinflammatory drugs (NSAID) intra muscular.

- After removal of milk push in suitable antibiotic onitment as per the advice of veterinarian.
- Always try to prevent, control mastitis than to treat the case.
- After calving (from 15th day to 200 days of calving)examine milk from each teat by simple cow-side C.M.T.(California Mastitis Test) once in 15 days. This will detect sub clinical mastitis in time and effective control measures can be taken. Mastitis is a man made disease, it needs proper hygenic measures to be adopted for udder care for control.

PARTURIENT HEMOGLOBINURIA BIOLOGY

Post parturient hemoglobinuria is a sporadic disease of multiparous, high producing dairy cows and buffaloes characterised by red blood cells breakdown in the muscles, hemoglobinuria and anemia. The prevalence of the disease in the overall cattle population is very little with a case fatality rate ranging from 10 to 50 per cent.

Parturient hemoglobinuria was formerly known as milk fever complex, post- parturient hemoglobinuria, puerperal hemoglobinuria and nutritional hemoglobinuria, etc.

HISTORY

Parturient hemoglobinuria was first reported in 1939 in buffaloes in the Indian sub-continent in the Lyallpur district, now Faisalabad. The first known reference of its occurrence and possible cause in Pakistan was by Hussain (1955). Subsequent reviews on the issue recommended that phosphorus insufficiency may be a possible reason of the disease. However, so far its true etiology remains doubtful.

ETIOLOGY

Consequent papers concluded that PPH is neither infectious nor contagious derived from negative serological and bacteriological verification for pathogenic bacteria and failure to identify erythrocyte parasites. Different causes, *e.g.* protein and mineral insufficiency, Saponin from cruciferous plants, competition for mineral assimilation and, hypophosphataemia because of phosphorus insufficiency.

Pirzada et al. (1989) and Cheema et al. (1980) reported that no contributory pathogen could be secluded from the affected animals. Copper insufficiency has been recommended as a possible cause of PPH by personnel in New Zealand. animals from dairy farms with a high occurrence of PPH had low levels of copper in serum and liver.

A disease in Ontario named as "red water" was linked with several predisposing factors which included: a) fresh parturition, b) intense milk production, c) dietetic phosphorus insufficiency and d) eating of turnips, rape, kale, green alfalfa and sugar beet pulp. In addition, many cows were hypophosphataemia.

SUSCEPTIBLE ANIMALS

A first round report recommended that parturient hemoglobinuria frequently affected high producing buffalo cows in the area of the Punjab province of Pakistan wherever there is a insufficiency of minerals, chiefly in the districts of Faisalabad, Jhelum, Attock and Rawalpindi. At these points soil tended to be deprived in minerals and buffaloes were affected more than cattle.

The incidence of this syndrome in bulls, cows prior to parturation, heifers under two years old, or beef cows is remarkable but has been reported. A disease with many similarities has been described in sheep, Egyptian and Indian buffaloes and a goat.

CLINICAL SIGNS

Clinical signs of syndrome can bee seen 20+,-10 days, before or after parturition at what time affected Animals pass red to coffee coloured urine and rectal temperature ranges from 38.38°C to 39.48°C. On the other hand, with the advancement of disease, rectal temperature declines. Additional clinical signs are jaundice, anemia, recumbency, in appetence, labored breathing, and constipation.

Cheema et al. (1980) reported that there was a considerable increase in the rectal temperature of affected animals. The disease affects buffaloes more than cattle which is a judgement in close agreement with that of Raz et al. (1988).

Pirzada et al. (1989) additional observed that phosphorus administration yielded effective outcome suggestive of hypophosphataemia as a causal issue and that, in the affected areas from August to January, animals in advanced stages of pregnancy should be supplemented with dicalcium phosphate or bone meal which should contain 13 per cent and 17 per cent phosphorus correspondingly. This agreed with the findings of Akram et al. (1990) and Raz et al. (1988).

Earlier, Sadiq et al. (1965) found that even though cases of parturient hemoglobinuria occurred all the way through the year, the occurrence increased in winter.

On the other hand, Cheema et al. (1980) reported that frequently stall-fed animals were affected, and that most cases were occurring in July and September.

The acute disease (three to five days) can come to an end in death or be followed by extended convalescence (two to eight weeks). Gangrene and sloughing of the extremities are reported sequelae. Recovered animals get back their former body condition and milk production gradually.

DIAGNOSIS

It can On the basis of clinical signs and history Urinalysis can be helpful in the diagnosis of this syndrome. Microscopic inspection of the urine sediment is imperative to distinguish hematuria from hemoglobinuria.

TREATMENT

The recommended treatment for Post parturient hemoglobinuria in North America includes:

- Intravenous infusion of sodium acid phosphate (60 g in 300 mL of water),
- 100 g of bone meal administered as a drench two times a day,
- Transfusion of fresh blood as needed and
- Intravenous fluids to sustain hydration.

Improvement of any phosphorus insufficiency or disproportion in the ration together with removal of incriminated feeds might prevent further cases. Because of incompatible results with phosphate therapy and the copper-deficient status of affected cows, personnel in New Zealand recommend parenteral copper (120 mg available copper per cow) as the favoured treatment.

A study on treatment trials of hemoglobinuria in buffalo cows and cattle indicated that subsequent treatment with sodium acid phosphate (20 per cent sol) administered concurrently by i.v., s.c. and oral routes, approximately 100 per cent of affected animals recovered in 1-3 days. Blood transfusion also shows a good result in this regard. A blood coagulant Botrophase prepared from the venom of the snake Bothrops jararaca, seems to have anti fibrinolytic action and was productively used to treat buffaloes facing parturient hemoglobinuria. Parenteral copper (120 mg available copper per cow) as the favoured treatment.

TREATMENT WITH TOLDIMFOS SODIUM AND TEA LEAVES AND SODIUM ACID PHOSPHATE

Result of this treatment was based on the recovery of urine discoloration; the efficacy of toldimfos sodium was 85 per cent followed by tea leaves 56 per cent, and sodium acid phosphate 18 per cent. Fallowing treatment with tae leaves and toldimofos sodium urine was clear next day, and with treatment with sodium acid phosphate urine was clear on third dy.

PREVENTION

Parenteral administration of copper has been effectual in dairy herds with previous histories of PPH (36,37). For prevention of the disease, supplementation with dicalcium phosphate has been suggested. The occurrence of PPH was considerably lower (5.18 per cent versus 25.51 per cent; $P < 0.01$)

in cows treated with copper previous to calving. moreover, a top dressing of copper sulfate to pastures four months previous to calving was followed by an increase in pasture, blood and liver concentrations of copper and a noticeable decrease in the occurrence of Post parturient hemoglobinuria. Pick up the soil phosphorus level by using phosphorus fertilizers for the reason that a study in India shows the connection between soil and this disease. One study which was conducted by Mohamed and El-Bagoury (1990), shows that there is a connection between berseem feeding and parturient hemoglobinuria for the reason that berseem is deficient in phosphorus thus avoid high quantity berseem feeding by feeding with some other fodders. Berseem also has Saponin which causes hemolysis causing hemoglobinuria.

POULTRY DISEASES

The physical environment afforded by a poultry research or teaching facility should not put birds at undue risk of injury or expose them to conditions that would be likely to cause unnecessary distress or disease. The facility should be maintained in such a way as to allow the birds to keep themselves clean and free from predators and parasites, prevent bird escape and entrapment, and avoid unnecessary accumu-lation of bird waste.

Environmental conditions are known to have major implications on the health, performance, and welfare of poultry. Air quality and the thermal environment should be maintained by ventilation, cooling, and heating to provide birds with the right environmental conditions for their age and time of the year. Welfare of the caretaker, in addition to bird wellbeing, deserves consideration in evaluation of housing systems and should receive attention during remodeling and development of future designs and concepts.

Bird exposure to high levels of ammonia causes irritation of the mucous membranes of the respiratory tract and eyes, increasing susceptibility to respiratory diseases. Birds detect and avoid atmospheric ammonia at or below 25 ppm. The *National Institute for Occupational Safety and Health* (NIOSH), the recommended exposure limits for humans should be no greater than 25 ppm for an 8-h day; for short-term exposure of 15 min, the threshold is 35 ppm.

Ideally, ammonia exposure for birds should be less than 25 ppm and should not exceed 50 ppm. Design of all housing systems should facilitate cleaning of the house and equipment as well as the inspection of birds.

Cages with multiple decks should allow for cleaning of equipment and inspection of birds without handling them, yet the birds should be easily accessible. Adequate lighting should be available for examination of all birds, and a movable platform or other system should be provided for examination of higher level decks, if those cannot be readily seen by attendants standing on the floor. Feeding and watering equipment also should be accessible for easy maintenance.

ADVANTAGES AND DISADVANTAGES OF CONVENTIONAL AND ALTERNATIVE HOUSING SYSTEMS

Although there are a variety of systems that can be used for housing poultry, including conventional and furnished cages, aviaries, littered floor systems, and free range, no housing system is perfect, with each system having its own health and welfare advantages and disadvantages. Research into alternative housing systems has been extensive in recent years including furnished cages, aviaries, and free-range systems as alternatives to conventional cages for egg-laying strains of chickens.

Conventional cages lack nests, perches, and dust baths to meet the behavioural needs of hens, but conventionally caged hens have less cannibalism and pecking because of smaller group sizes leading to a reduced trend in mortality compared with hens in non-cage systems. Because conventional cages lack perches and do not have access to litter, poor foot health and keel bone deviations and deformities are not as problematic in cages as they are in non-cage systems or furnished cages; however, because of lack of exercise, conventionally caged hens are susceptible to osteoporosis.

Moreover, freerange birds are able to express behaviours such as freedom of movement, running, short-distance flying, and the scratching of soil, and have the opportunity to be exposed to a variety of environmental stimuli. They are also leaner with more muscle mass and plumage than caged birds.

However, ranged birds are more susceptible to problems caused by inclement weather and have increased risks of bacterial disease, parasites, cannibalism due to larger group sizes, predators, environmental contaminants such as dioxin, and increased frequency of old bone fractures. No housing or management system is likely to be ideal in all respects. Therefore, ethically acceptable levels of welfare can exist in a variety of housing systems.

ALTERNATIVE HOUSING

FURNISHED CAGES FOR EGG-LAYING STRAINS OF CHICKENS

Furnished cages are available to house large, medium, and small group sizes. The European Commission offers standards for furnished cages that include perching space for all hens and a nest and dust bath area, with minimum available space per hen of 750 cm^2 per bird.

Appleby suggests that group sizes of 8 hens or more in furnished cages should have 800 cm^2/hen and that smaller groups of 3 or less should have 900 cm^2/hen, plus an area with litter. In these systems, claw-shortening devices are helpful to maintain short claws, and perches can help to increase leg strength. Problems observed in this type of housing include increased keel bone deformities associated with high perch use and should be monitored.

AVIARIES OR MULTI-TIER SYSTEMS FOR EGG-LAYING STRAINS OF CHICKENS

Aviaries, designed to use vertical space, consist of a ground floor plus one or more tiers consisting of perforated or slatted floors or platforms with manure belts underneath. Providing a littered area allows for dust bathing and reduces the incidence of cannibalism and feather pecking. The scratch area also allows the hens to keep their claws trimmed. The litter should cover enough area to allow for proper mixing of manure and avoid excessive manure and moisture accumulation.

The depth of the litter should be sufficient to prevent hens from coming in contact with the floor. Likewise, the depth of the litter should not be so deep that it encourages the laying of eggs on the floor. Opening and closing the littered areas for specified periods can be used as a management tool to prevent the laying of floor eggs. The European Commission recommends that the littered areas cover at least 30 per cent of the useable floor area of the house. The recommended floor space per hen for aviaries excludes nest space. Only the floor area and the tiers can be counted as usable space when calculating stocking density for hens in aviaries. Hens housed in aviaries have a high incidence of bone fractures during the laying cycle because of crash landings or failing to jump gaps effectively. Each tier should allow hens to safely access other vertical tiers, including the littered floor. For example, a ramp can be used to allow birds to move from the littered floor area to the first raised tier.

If ramps are used, they should be designed to prevent droppings from falling on the birds below. Hens should have access to the entire littered floor area, including the area under the raised tiers. Raised tiers need a system for frequent removal of manure. To reduce the incidence of hen injury, including broken bones, the highest tier should not exceed 2 m.

Vertical distance between tiers, which also includes the floor to the first tier, is recommended to be between 0.5 and 1.0 m. Measurements may be taken from the top of the littered floor or slat area to the underside of the manure belt. When adjacent tiers are staggered to allow for diagonal access to tiers of different heights, the hen's angle of descent should not exceed 45°.

The horizontal distance between tiers should not be more than 0.8 m. Where design discourages horizontal movement between different tiers, there should be a minimum distance between tiers of 2 m. For flock sizes that exceed 3,000 hens in a room, no more than 2 raised tiers above the floor are recommended. Smaller flock sizes of 3,000 or less can have up to 3 raised tiers in a room. Birds that are to be housed in aviaries as adults should be reared as pullets in similar aviaries to facilitate adaptation to perches and nests. Typically, day-old chicks are housed in a central tier the first 10 d of age and then about half of the pullets can be distributed to the lower tier to provide more space as they age.

In this manner, the pullets quickly find the feed and water and are provided proper brooding temperatures during the early stages of growth. By 15 to 21 d of age, pullets are given full access to the aviary. Ramps are provided to allow pullets easy access to all levels of the aviary. Perch space per pullet is recommended to be 8 cm/pullet during the first 10 wk of age and 11 cm/pullet after 10 wk of age. Welfare standards for pullet aviaries are still in the investigational stage.

OUTDOOR ACCESS OR FREE RANGE

Poultry may also be raised with access to the outdoors. Poultry raised under an organic protocol require outdoor access, which can be a range or a semi-enclosed yard often referred to as a veranda or winter garden. During inclement weather or for health-related reasons, birds should remain indoors or in shelters until such conditions are improved. A range is an outside fenced area. Fence height and fencing material should be of appropriate mesh size to retain domesticated poultry and prevent predator entry.

A permanent fence can be extended underground to a minimum depth of 0.25 m to prevent ground predator entry. The fence can be surrounded by an electric wire 25 to 45 cm above the ground and 0.6 to 1.0 m away from the primary fence. Overhead fine netting, as used for game birds, can be used to protect domestic poultry from wild avian predators and minimize disease transmission from wild species to domesticated poultry.

Ranges should be free of debris such as large rocks and fallen trees, environmental contaminants, and be designed to prevent muddy areas, to avoid injuries and foot problems, and to promote overall bird health. Vegetation should be used for ranges or sections of the range where soil erosion is problematic. Range rotation is one tool for minimizing the threat of a disease outbreak and to provide opportunity for land to recover from bird activity.

A covered veranda provides shade and is connected to the house and is made available to the hens during the daylight hours. The floor of the veranda can be solid and may be covered with litter. To minimize the probability of cannibalism, natural light or high-intensity artificial light can be used during early stages of rearing to facilitate the transition of birds from indoor to outdoor lighting conditions. Free-ranged birds without access to a permanent building should have covered shelters that provide shade, protection from inclement weather, litter, food, and water. The sheltered area should provide space to allow all ranged birds to rest together without risk of heat stress. Mobile shelters should be moved on a regular basis or managed to minimize the probability of a disease outbreak or muddy conditions. Elevated perches designed for poultry can be provided on the range or inside the indoor shelter. All range, veranda, or any other type of outdoor access should be managed so that birds are protected from potential predators.

Weather permitting, birds should be given access to the outside as soon as they have full feather coverage to encourage ranging behaviour.

Vegetation such as small bushes, crops such as corn, or cover panels that provide a sense of protection in the outdoor area can be used to encourage the use of the range. When indoor birds are allowed free access to the outdoors, they should have appropriately sized openings of sufficient number to facilitate bird exit from and entrance into the building; alternatively, the doors of the house can be opened to allow birds freedom of movement.

The size of each pophole should allow for easy passage of a bird to and from the outside. The number of popholes provided should allow birds to comfortably access the outside or inside without significant congregation of birds on either side of the pophole. A roof can be placed over a pophole to provide protection and baffles installed to reduce entry of wind into the house.

Slats can also be used to prevent the formation of muddy areas around the popholes. For whole house configuration without individual pens, popholes should be evenly distributed down the entire length of the building to prevent birds from blocking the access in and out of the building. On windy days, it may be wise to open popholes only on the leeward side, so providing more than the minimum number of popholes is advisable.

EGG-LAYING STRAINS OF CHICKENS

The approximate age that egg-laying strains of chickens are allowed access to the range is about 12 wk of age. Before 12 wk of age, they are brooded in confinement. To allow for range rotation, provide each hen with 4 m^2 of outdoor access. Shade should be evenly distributed in the outdoor area and provided at a minimum of 8 m^2 per 1,000 hens.

MEAT-TYPE CHICKENS

Fast-growing strains of broilers should have access to a minimum of 1 m^2 of outdoor access, whereas slower growing strains require 2 m^2 of outdoor access.

TURKEYS

The age that turkeys are given access to outdoors may vary from 5 to 12 wk depending on weather conditions and predator risk, with 8 wk being the most common age. A flock can gradually be transitioned to range by moving one-third of the flock the first morning and then moving the remainder of the flock a day or two later. The following formula can be used to calculate the minimum amount of shelter recommended: area, m^2 = [W]/D, where n is the number of birds in the flock, W is the expected average weight at depopulation, and D is the maximum stocking density in kg/m^2. Growing turkeys are allowed a minimum space allocation of 6 m^2/bird of free range.

DUCKS

Information for porches or winter gardens for ducks is not available. When growing ducks are first introduced to the range, they need to be shown the location of the feeders, drinkers, and shelters. The outdoor feeders and drinkers should be surrounded by slatted or solid flooring to prevent the ground in the immediate area from becoming muddy.

Free-ranged growing ducks are allowed a minimum of 2.5 m^2/bird when reared on well-maintained ranges with ground cover. If the vegetation is poor, then a minimum of 4 m^2/growing duck should be provided. If ponds are available, they should be well maintained so as to avoid stagnant water containing decaying vegetation. Botulism in ducks can be a problem when pond water is not well aerated or not filtered to remove plant debris. Developing breeders may be raised outdoors on well-drained soil with open shelter. A minimum of 1,290 cm^2 of shelter area/bird is recommended for developing breeders.

LIGHTING FOR ALTERNATIVE POULTRY PRODUCTION

Energy-Efficient Lighting for the Farm from the US National Sustainable Agriculture Information Service (ATTRA) provides an overview of energy-efficient lighting technology and explains how to select lighting options that are appropriate for the farm, including alternative poultry production. Energy-efficient lighting options present farmers with new opportunities to reduce electricity costs and help manage farms sustainably. Cost-effective energy-efficient lighting can be used to improve productivity and safety, and reduce operating costs.

Supplemental lighting is normally used by alternative egg producers to maintain productivity, and sometimes for alternative broiler production in northern climates. Small layer flocks housed during late spring through mid-summer with daily access to the outdoors do not require supplemental light. Supplemental lighting is necessary for pullets to maintain production during late fall and winter as days shorten. Poultry are very sensitive to three aspects of light: intensity of light (measured in footcandles), wavelength (measured in colour temperature), and day length (duration of light period). Research by Michael Darre and others has found that blue light wavelengths help calm birds; red wavelengths may be used to help reduce feather picking; blue-green wavelengths help maintain growth; and orange-red wavelength helps maintain reproduction.

The light intensity for layers should be enough to read a newspaper by and will vary with the poultry breed. Generally, 'warm' wavelength lamps of less than 3,000K in the red-orange spectrum are best for small flocks with outdoor access. The day length should never be extended past 16 hours or the longest day of the year. Solar photovoltaic lighting provides a simple solution to maintaining egg production during shorter days. Solar lighting systems basically consist of a solar module, a deep-cycle battery, a charge controller, a 12V programmable timer, and an efficient DC lighting fixture with lamp. Energy-efficient LED lamps work very well with solar modules. All of the components to build a basic low-voltage solar lighting system can be purchased online for less than $300 or as a kit.

To conserve energy and keep poultry healthy, use timers to switch lights on and off. Programmable timers must be 12V when used in conjunction with a 12V solar lighting system. Th ere are 12V timers available online as well as schematics to convert a household programmable thermostat to a 12V timer. Timers also ensure that birds receive a uniform number of light hours each day. Set timers to light in the morning instead of the evening to give birds a natural dusk and allow them to roost. Check timers at least once a week, and clean lamps if dust builds up. Lamps should be free of obstructions that cause shadows on the floor. Baby chicks require additional light in their first 72 hours to help them find food and water. A low watt 'warm' lamp is recommended for every 200 square feet of floor space. (Hawes) The high heat from incandescent lamps may double as a brood light and heat source, although it may be more energy-efficient (and cost-effective) to use a separate heat source and a solar lighting system.

ALTERNATE POULTRY PRODUCTION IN INDIA

The alternate poultry production in India is gaining momentum and attention from the farmers, entrepreneurs, professionals and researchers. It is used for food, game, fancy, pet, and also for research purposes. Here, the importance, prospects and potentials of alternate poultry production in India has been reviewed.

QUAIL PRODUCTION

Quail is the most efficient biological machine for converting feed into animal protein of high biological value and hence is the cheapest source of animal protein for human diet. Consumption of quail began before domestication of poultry. In China, Taiwan, UK quail has been reared through centuries. During fifties its commercial production started in Italy. The CARI is the pioneer in introducing quail farming in 1974 by importing germplasm from U.S.A.

It is introduced in India as an alternate avian species for growing poultry industry. M/s. AVM Hatchery and Poultry Breeding Research Centre, Coimbatore has played a major role for production on extensive scale in private sector. Owing to the wild life act and ban on captivation of domestic quail, no worthy development and popularisation of quail production took place for last ten years. But now the ban has been lifted, it has tremendous scope to occupy a considerable portion of poultry meat market by 2010 A.D. as a profitable enterprise. Quail has unique qualities of hardiness and adaptability to diversified agro-climatic conditions. Its consumption may be referred by all, as it has no religious taboo. It has nutritive value, amasing taste, and game favour, tender, very delicious with low calorific value and high dry matter.

It is rich in protein, vitamins, essential amino acid, unsaturated fatty acids and saturated fatty acids and phospholipids. It may be recommended for children and pregnant women, for speedy recovery. It is a tiny inter breeding avian species quite different from wild quail. Because of twin capacity of prolific egg production and meat yield, it attains the status of viable commercial poultry enterprises as well as rural poultry production.

At present quail production is being done at Agricultural Universities or few private sectors only, but if we want to achieve our target we will have to extend our quail farming activities to rural areas where our bulk of population lives. Quail production requires less investment and gives quick returns, higher profits and hence can be adopted by rural masses quickly. A very important point is that so far benefits of quail farming have not fully reached rural masses, though quail production can generate additional income. Agriculture production can be boosted through the use of high value quail manure, as it is four times nutritive than cow dung manure.

DUCK PRODUCTION

Ducks are the second largest source of table eggs and there are about 23 million ducks in India. West Bengal has the highest duck population followed by Assam, Tamil Nadu, Andhra Pradesh, Bihar, Kerala, Orissa, Jammu and Kashmir and Tripura. Ducks predominantly are of indigenous type kept for egg production on natural foraging and have a production potential of about 130-140 eggs/bird/year. Duck eggs have a preference over chicken eggs in certain states and areas.Considering the importance and scope of duck rearing in our rural economy, improved strains and technologies are available in the country and the farmers are much benefited from the same.

Duck farming is primarily popular among small farmers, marginal farmers and agricultural labourers as well as poor section of the community. Ducks lays 40-50 eggs more than desi chicken. The duck egg is heavier than hen by 15-20g. Ducks require lesser attention and supplement their feed by foraging, eating fallen grains in harvested paddy fields, insects, snails, earthworms, small fishes and other aquatic materials in lakes and ponds hence incur reasonable feed cost.

Ducks have a longer profitable life and lay well even in second year. Ducks do not require elaborate houses like chicken, reducing the capital investment. Ducks are quite hardly, more easily brooded and more resistant to common avian diseases. Ducks flourish well in marshy riverside, wetland and barren moors where other types of stocks do not. Ducks have no problems of cannibalism or pugnacious behaviour. Ducks lay 95-98 per cent of their eggs early in the morning, before 9 AM, thus saves a lot of time and labour. Ducks are useful in controlling unwanted plants in ponds, lakes, and streams like green algae, duckweed, pond weed, musk grass, arrow head wild celery, etc. (5-10 ducks/ 0.405ha of water table). Ducks are good exterminators of potato beetles, grasshoppers, snails, and slugs. In areas plagued by liver flukes ducks can help correct the problem (2-6 ducks/ 0.405 ha of land). Ducks can be used to free the bodies of water form mosquito pupae and larvae (6-10 ducks/ 0.405 ha of water space).

Ducks are suitable for integrated farming systems such as duck-cum-fish farming, duck farming with rice culture, etc. In duck-cum-fish farming the droppings of ducks serve as feed for the fishes and no other feed manuring of the pond is necessary for fishes (200-300 ducks/ha of water area). Under integrated duck farming with rice culture, the ducks perform 4 essential functions, *viz.*, intertillage, weeding, insect control and manuring. The down and small body feathers of ducks are valuable as filler for pillows and as lining for comforters and winter clothing (one Pekin duck produce about 100g of down and small feathers).

The important Indian breeds are Sythet mete and Maheswari, of eastern region, Chara and Chamballi of Kerala, desi breeds of West Bengal and other states. Among improved varieties distinct breeds for egg and meat production are available. Khaki Campbell and Indian Runner are the most popular breeds for egg laying. White Pekin, Muscovy and Aylesbury are known for meat production. High-laying strains of ducks are available at the Central Duck Breeding Farm, Hessarghatta, and Bangalore. Duck units are functioning in some of the states like Assam, Tripura, Orissa, Madhya Pradesh, Andhra Pradesh, Kerala, Haryana and J & K.

TURKEY PRODUCTION

Turkey is mainly reared for meat and is quite popular in many western countries. It is one of the white meat choice and famous for its leanness and deliciously cooked whole turkey forms the traditional dish in Thanks giving and Christmas dinners. Turkeys are sold as whole or cut-up parts and further processed products such as sausages, patties, and nuggets.

Turkey farming is still at the stage of infancy and need to be popularised among farmers to provide diversified food and employment especially in India. Rearing turkeys can be an excellent family or youth project. They are quite suitable for upliftment of small and marginal farmers as the birds can be easily reared in free range or semi-intensive system with minimal investment for housing, equipment and management.

Even though turkey farming is practically non-existent in our country, few backyard birds are distributed in few areas. Also organised farms of turkeys have been maintained at Central Avian Research Institute, Izatnagar, CCS Haryana Agricultural University, Hissar, University of Agricultural Sciences,

Bangalore, Rajasthan College of Agriculture, Udaipur, Kerala Agricultural University, Mannuthy and Tamil Nadu Veterinary and Animal Sciences University, Kattupakkam.

Turkeys are not difficult to raise. They require a little special care to get them off to a good start. Sometimes they are little slow in learning to eat and drink. If the farming is started with a good stock and provided optimum feed, housing and management, one can raise turkeys successfully.

There are seven standard breeds of turkeys, *viz.* Large White (Broad Breasted White), Bronze (Broad Breasted Bronze), Beltsville Small White, Bourban Red, Black, Slate and Narrgansett. The turkeys can be marketed as broiler at approximately 16 weeks of age when hens usually reach a live weight of about 8kg. Toms weigh approximately 12kg. Smaller, fryer roasters can be produced by slaughtering at an earlier age as per the market demand. Adequate research in the field of management, nutrition and disease control under tropical conditions as well as education are imperative before private entrepreneurs can be convinced for the viability of turkey enterprise in India.

OSTRICH AND EMU PRODUCTION

Ostrich and Emu farming is now-a-days considered to be one of the most profitable agricultural projects. They are referred as 'farms of the future' due to the large variety of products and hence the high profits potential. Ostrich meat is highly nutritive and earns enormous foreign exchange. Ostriches are raised commercially for their egg, meat, feathers and hide. The increased consumer awareness of the problems of cholesterol in the blood and their possible association of heart attacks and arteriosclerosis, the demand for ostrich meat in international market began to increase. A number of beef producers in Europe, America and Canada have most recently switched to commercial ostrich farming due to higher and faster financial returns on ostrich projects. Ostrich meat is in great demand in European and American countries and is already extremely popular and available in most of the restaurants and supermarkets.

Recently Tamil Nadu Veterinary and Animal Sciences University imported 100 chicks from Malaysia. Their adaptability to climatic conditions prevailing in the state would be tested and if successful it will be recommended to the interested farmers and entrepreneurs. Considering the above aspects of alternate poultry production, our country has a great potential to develop further in this area of farm production. The NABARD and associated banks are also ready with model projects and financing of such projects through reimbursement scheme which is also a boost for future developments.

GUINEA FOWL PRODUCTION

An important source of traditionally raised poultry species, guinea fowl was introduced into Indian sub-continent by African ethnic migrants a few centuries back. Definite population figures are not available, it is estimated that few million birds are raised every year. Guinea fowl ranks third among the poultry birds; it is referred by different local names in different regions, *viz.*, chittra in west, china murghi in south and titari in the northern plains. Earlier attempts to raise guinea fowl intensively were made at Christian Missionary Poultry Farm, Etah (UP), Allahabad Agriculture Institute, Naini (UP), and Government Poultry Farm, Gurdaspur (Punjab) and subsequently at Military Poultry Farm.

However preliminary survey of guinea fowl production in northern regions and the perusal of earlier published literature established its distinct popularity with marginal farmers, tribes and pastoralists may be attributed to its inherent hardiness and excellent foraging potentials to go as animal component of biomass-based polycultural systems.

In recent years this alternate poultry species witnessed increasing emphasis for low-input grain-saving aviculture (LISA).

The primary function of guinea fowl production is to produce low-cost game flavoured meat. There are two quite distinct production systems for rising of guinea fowl. One is the modern intensive system and other is the traditional extensive system.

The former is characterised by high inputs, high costs, good husbandry, hygiene and a complete supply of feed. The intensive production is largely confined to Government and public sector owned farms. Commercial farming is not much known. Traditional extensive system is the most common production system followed in rural area.

Guinea fowl holds a unique status among the other poultry species owing to its low input requirements and better forage-utilisation capacity. There are three major varieties of guinea fowl, *i.e.* Pearl, Leavender and White.

Guinea fowl deserve priority research attention for helping the smallholder rural poultry production. Research and development activities should aim not primarily at designing completely new technologies to replace the traditional ones but rather understanding and strengthening them on social, ecological and economic merits.

FEED AND WATER

FEED

Circular or linear troughs can be used to supply feed. Feed troughs can be located either inside or outside the area where the birds are housed. If feed troughs are located outside the area where the birds are housed, then only one side of the trough is available to the birds. Unless the feeder is mounted on a wall, feeders located in the area where the birds are housed generally provide bird access to both sides of the trough.

Depending on species, specifications are for birds housed in multiple-bird pens and cages, individual cages, or aviaries. Feeder space allocation is presented as linear trough space per bird when both sides of the trough are available. If only one side of the trough is available, then the amount of feeder space per bird must be doubled. Because meat-type chickens, ducks, and turkeys have been bred for rapid growth to market age, excessive *body weight* (BW) gain of broiler breeders, duck breeders, and male turkey breeder stocks is a problem unless energy intake is controlled beginning early in life.

Because breeders are allocated limited feed to allow for a gradual increase in BW each week, birds are hungry as indicated by motivational test, stereotypic pecking on non-nutritive objects, and excessive drinking of water. Stress is also apparent in feedrestricted broiler breeders between 8 and 16 wk of age. Feed restriction of breeders allows for controlled BW gain, reduces skeletal problems, increases activity, and improves livability, fertility, immune function, egg production, and disease resistance. Evidence to date indicates that the welfare of breeders is better if they are feed restricted.

Feed should be allocated and BW routinely monitored to maintain the recommended BW for the particular stock and age. Rations may be either a fixed amount of feed allotted daily or under various alternate-day feeding schemes. Alternate-day feed restriction as opposed to limited feed each day allows more-timid birds access to feed, resulting in better flock uniformity. Inhibition of feeding by subordinate birds is likely if feeder space is limited. Therefore, procedures that require restricted feeding should have enough feeder space so that all birds can eat concurrently. It may also be helpful to use low-density diets and to provide birds with environmental enrichment such as devices that they can manipulate to obtain small amounts of food to fulfill their feeding behaviour. Although adult broiler breeders are housed together for mating, they are fed separately to control BW gains.

If both sexes have access to the same feeder, the more aggressive males will consume more than their share of feed. The female feeder is fitted with a 4.3-cm grill sufficiently wide to allow feeding, whereas the male trough is fitted with a 5.1-cm grill. In this manner, the installation of narrow grills over the female feeder may prevent males with larger heads from consuming the hen's feed.

However, some genetic lines of male breeders have smaller heads allowing them access to the female feeder, which not only deprives the hens of proper nutrient intake, but may lead to excessive BW gains for those males eating the hen's feed. University research uses a multitude of genetic lines in their studies; therefore, a one-size restriction grill does not exist to meet the head size of all breeds of meat-type chickens. To rectify this situation, small plastic pegs that are 6.3 cm in length are inserted through the nares of genetic lines of male broiler breeders known to have small heads at 20 to 21 wk of age to minimize male access to female feeders. The behaviour of males with Noz-Bonz inserted did not appear to be affected, with resumption of foraging activities immediately post-insertion. Use of breeds or genetic lines that do not require Noz-Bonz is highly encouraged. Ducks experience difficulty consuming mash because the mash, as it becomes moist, may cake on their mouth parts. Therefore, it is recommended that all feeds for ducks be provided in pelleted form. Pellets no larger than 0.40 cm in diameter and approximately 0.80 cm in length should be fed to ducklings less than 2 wk of age. Pellets 0.48 cm in diameter are suitable for ducks over 2 wk of age.

WATER

Recommendations for watering space vary widely, depending on species, type of bird, bird density, and whether water intake is restricted. Depending on type of poultry, specifications are for multiple-bird pens and cages, individual cages, or aviaries. These recommendations assume moderate ambient temperatures. Newly hatched birds may have difficulty initially obtaining water unless they can find the waterers easily. Similar difficulties may occur when older birds are moved to a new environment, especially if the type of watering device differs from that used previously by the birds.

Watering cups that require birds to press a lever or other releasing mechanism involve operant conditioning. Because individuals may fail to operate the releasing mechanism by spontaneous trial and error, shaping of the behaviour may be required. Thus, it may be necessary to press the individual bird's beak or bill to the trigger to facilitate finding the water source.

Watering cups may need to be filled manually for several days until the birds have learned the process. Water pressure must be regulated carefully with some automatic devices and watering cups. In such cases, pressure regulators and pressure meters should be located close to the levels at which water is being delivered. Manufacturer recommendations should be used initially and adjusted if necessary to obtain optimal results.

Automatic watering devices require frequent inspection to avoid malfunctions that can result in flooding or stoppage. Waterers should be examined at least once per day to ensure they are in good working condition. The height of drinkers should be adjusted to meet bird size. Birds accessing nipple drinkers should raise their heads up while standing to activate the trigger pins. As a general guide, the bottom of the water trough should be approximately even with the back of the bird. Poultry ordinarily should have continuous access to clean drinking water. However, with some restricted feeding programmes, overconsumption of water may occur, leading to overly wet droppings that can hamper health and performance of poultry due to poor litter quality.

This situation can be controlled by restricting excessive water intake, usually by limiting water availability to certain times of the day, in accordance with accepted management programmes that consider the amount of time that feed is available and also environmental temperature conditions. There is little effect on welfare indicators of breeders with limited access to water compared with breeders consuming water ad libitum.

Water should be provided each day and also made available during the time that feed is being consumed. Adequate drinker space is needed to prevent undue competition at the drinkers when the water is turned back on. Water may also be shut off temporarily in preparation for the administration of vaccines or medications in the water. Most conventional poultry drinkers may be used for ducks, except for cup drinkers that are smaller in diameter than the width of the duck's bill. Nipple drinkers support slightly poorer duck performance during hot weather than do trough waterers. Ducks can grow, feather, and reproduce normally without access to water for swimming or wading, but weight gain may be improved slightly during summer months if such water is provided. If ducks are provided water for swimming or some other wet environment, they should also have access to a clean and dry place; otherwise, they are unable to pre-en their feathers and down properly, and the protection normally provided by this waterproof, insulated layer may be lost.

HUSBANDRY

SOCIAL ENVIRONMENT

All poultry species are highly social and should be maintained in groups when possible. However, certain social environments can be stressful to poultry and should be avoided. For example, repeated movement of individuals from one socially organized flock to another may induce stress in those individuals that are moved. Human interactions with chickens can also contribute, either favourably or unfavourably, to the social environment of the animal.

A calm, friendly interaction between known animal caretakers and the birds will result in reduced stress and better performance compared with abrupt, careless interactions. Chickens, turkeys, and ducks are likely to panic when sudden changes occur in their environment. When birds are kept in group housing, this panic reaction may result in birds trampling each other and piling up against barriers or in corners with resulting injury and mortality. Husbandry methods should be used to prevent death loss caused by smothering. Such sudden changes should be prevented to the extent possible. Alternatively, young birds, which are less reactive to such stimuli, can be habituated to conditions that are likely to be encountered and could cause panic responses later in life.

CHICKENS

Excessive fighting and mounting may occur in groups of mature males residing in floor pens. If such abuse is likely to be encountered, as when aggressive stocks are used, late adolescent or mature males should be placed in environments where those behaviours are not possible or are less injurious; for example, in individual cages, in multiple-bird cages with moderate density or in mixed-sex flocks with appropriate sex ratios.

The proportion of mature males in sexually mature flocks should be low enough to prevent injury to females from excessive mounting. Male to female ratios for breeding purposes can be variable in regard to different breeds and strains of chickens. The optimal ratio in most breeder flocks is 1 male to 12 to 15 females for egg-type strains and 1 male to 9 to 11 females for meat-type chickens. Some environmental enrichment techniques can be used to control aggression and over-mating in poultry. Recent research has shown that social dynamics in layers and chickens raised for meat are complex and increments in group size or density do not necessarily result in a linear increase in aggression or reduced welfare and performance. Intermediate group sizes of around 30 birds were found to be more problematic than smaller or larger groups of layers in floor pens. Chickens kept for meat production can be safely maintained in large groups of several hundreds or thousands of birds with no increased aggression or behavioural problems, as long as sufficient feeding and drinking space is provided to prevent competition for resources. The welfare of broiler chickens tends to be affected more by environmental conditions than by group size or density effects, as long as density is maintained within a reasonable range.

TURKEYS

Tom turkeys are prone to excessive aggression as they become older. Early beak trimming reduces the likelihood of injuries from fighting among toms. Breeder toms are housed separately from breeder hens using artificial insemination to produce fertile hatching eggs.

DUCKS

Ducks, being very sociable animals, do not perform well in isolation. Therefore, it is imperative that individually caged ducks have some means of social interaction such as a wire partition between adjacent cages so that they can see and touch each other. For sexually mature breeder ducks, injury to females resulting from excessive mounting by drakes may be exacerbated in the presence of other stressful conditions such as lameness associated with foot pad trauma caused by improper flooring. For Pekin breeders, the ratio of males to females should not exceed 1:5 and may require periodic adjustment throughout the breeding cycle because of higher mortality rates for females than for males.

FLOOR AREA AND SPACE UTILIZATION

Chickens, turkeys, broilers, and ducks should have sufficient freedom of movement to be able to turn around, get up, lie down, and groom themselves.

Use of floor area by birds within groups follows a diurnal pattern and is influenced by the dimensions and design of the facilities. Birds may huddle together for shared warmth or spread out for heat dissipation. They generally use less area during resting and grooming than during more active periods and will often seek the protection offered by the walls of the enclosure.

Floor space allowances for layer-type chickens in conventional cages are based on extensive research. In a survey of experiments involving density effects, Adams and Craig made multiple comparisons within specific categories for several production traits and for livability. Their survey indicated that livability and hen-housed egg production were reduced significantly when areas of 387 cm^2 and 310 cm^2 were compared with 516 cm^2, amounting to reductions of 2.8 and 5.3 per cent in livability and 7.8 and 15.8 eggs per hen housed, respectively.

Decreases in livability and other measures of wellbeing were also associated with high density. Craig *et al.* found that livability and egg mass were significantly lower with 310 cm2 than with 464 cm^2; Okpokho *et al.* and Craig and Milliken found livability was lower at 348 cm^2 than at 464 cm^2 and 580 cm^2; and Craig and Milliken found lower hen-day rate of lay and egg mass per hen at the highest density. In the same studies, however, no differences in survival and egg production measures were detected between the 2 lower densities.

From data on plasma corticosterone concentrations, Mashaly *et al.* concluded that more than 387 cm^2 of space per hen should be provided; Craig *et al.* found that plasma corticosterone concentrations were greater at 310 cm^2 than at 464 cm^2.

Similarly, feather condition was worse and fearfulness was greater when estimated at 40 wk of age or older. Using data on egg production, mortality, and serum corticosterone concen-trations, Roush *et al.* concluded that 3 hens, rather than 4, should be kept in cages of 1,549 cm^2 area; that is, within the goals and constraints employed, hens should have 516 cm^2 rather than 387 cm^2 area. Using operant determination for laying hens' preference for cage size, Faure indicated that a stocking density of 400 cm^2 was sufficient most of the time, although hens would work to obtain more space up to 25 per cent of the day. Modification of commercial cages from those currently in wide usage for chickens may improve the health and welfare of birds. Thus, cage height should allow birds to stand comfortably without hitting their heads on the top of the cages.

Studies have indicated at least 40 cm over 65 per cent of the cage area and not less than 35 cm at any point is desirable. Taller cages may be necessary for larger breeds. Cage floors with a slope of no more than 9° in shallow, reversed cages may result in better foot health. However, such low slopes may not be desirable in deeper cages, because difficulties are encountered in getting eggs to roll out efficiently. Horizontal bars across the front of the cage appear to allow egg-laying strains of chickens to feed easily and with reduced probability of entrapment.

White Leghorn hens housed in cages with horizontal cage fronts had better feather scores than hens in cages with vertical bars fronts. The cage door should be wide enough to allow easy removal of the bird. Caged hens may cease egg production temporarily or birds may undergo a molt if removed from the cages to which they have become accustomed; for example, for cage cleaning.

Therefore, hens and roosters may be kept in their cages for 18 mo or longer, as long as air cleanliness is maintained and excreta are disposed of regularly from under the cages. However, the incidence of osteoporosis and weak bones may be higher in hens caged for prolonged periods compared with hens housed in systems where greater freedom of movement is possible.

The welfare of meat chickens is not compromised at densities of 15 to 17 birds/m^2 as long as adequate environmental conditions are maintained. However, welfare status for a given density will depend in part on the final BW at which the birds are grown and managed. For example, heavy male broilers raised to 49 d of age at a stocking density of 30 kg of BW/m^2 had the lowest incidence of foot-pad lesions, the lowest incidence of scratches on the back and thigh, and had the best market BW compared with higher stocking densities of 35, 40, and 45 kg of BW/m^2.

With 35 d-old male broilers grown to a lower BW of 1.8 kg, feed consumption, feed conversion, BW gain, and foot pad lesions were adversely affected with increasing stocking densities. These results on lighter weight broilers suggested that the best bird performance and welfare was achieved at 25 kg of BW/m^2. In terms of space use, there is no scientific evidence to suggest that social restriction on use of space occurs in large groups of broilers, even in mature broiler breeders.

Although less active than layer strains, meat chickens will use more space when available to them. Studies have also shown that provision of partitions such as cover panels help to maintain a more-even bird distribution in the facility and can help to control behavioural problems. Use of space can be improved by providing rectangular rather than square pens for the same available area. Although broiler chickens can be maintained in cages, it is best for their health and welfare to use floor pens provided with some type of litter such as wood shavings. Because of a relative absence of research on well-being indicators for turkeys and ducks, recommendations are based on professional judgement and experience. Generally, area allowances are assumed to be adequate when productivity of the individual birds is optimal and conditions that are likely to produce injury and disease are minimal.

Singly caged birds are frequently used in agricultural research and teaching to establish or demonstrate fundamental principles and techniques. Because withincage competition for feed and water is absent, feeding and watering spaces are not critical; however, individually caged birds must have ready access to sources of feed and water except during feed-restriction periods for meat-type breeder birds.

FLOORING

Poultry may be kept on either solid floors with litter or in cages or pens with raised wire floors of appropriate gauge and mesh dimension. When poultry reside on solid floors, which are more adequate for heavy strains of poultry, litter provides a cushion during motor activity and resting and absorbs water from droppings. The ideal litter can absorb large quantities of water and also release it quickly to promote rapid drying. A dry, dusty litter or a litter that is too wet will have a negative effect on the health, welfare, and performance of poultry. Litter, when sampled away from the drinkers, needs to be moist but not so moist that it forms into a ball when handled. Litter should not emit excessive dust when disturbed. The poultry house should be ventilated to maintain litter in a slightly moist condition. Avoiding excess moisture in the litter improves bird health by reducing dirty foot pads, hock lesions, leg defects, and fecal corticosterone.

Some examples of acceptable materials used for litter, depending on local availability, include rice hulls, straw, wood sawdust or shavings, and cane bagasse. Because litter materials differ in their ability to absorb and release water, husbandry practices should be varied to maintain proper litter conditions. Litter being stored for future use should be kept dry to retard mold growth. When poultry are kept in cages or on raised floors, accumulated droppings should not be permitted to reach the birds. Droppings should be removed at intervals frequent enough to keep ammonia and odours to a minimum.

DUCKS

Particular attention should be paid to the type of floor provided in pens or cages for the common duck because the epidermis of the relatively smooth skin on the feet and legs of this species is less cornified than that of domesticated land fowl and, therefore, is more susceptible to injury. Properly designed, non-irritating floor surfaces minimize or prevent injury to the foot pad and hock and minimize subsequent joint infection.

Dry litter floors are least irritating to the feet and hock joints of ducks and should be used whenever possible, particularly if ducks are going to be kept for extended periods. Litter floors that are not kept dry present a serious threat to the health of the flock. Wire floors and cage bottoms of proper design may be used without serious adverse effects if the ducks are not kept on wire for more than 3 mo.

Younger ducks and smaller egg-type breeds are less susceptible to irritation from wire than are older and larger meat-type breeds. Properly constructed wire floors and cage bottoms should provide a smooth, rigid surface that is free of sags and abrasive spots. The 2. 5-cm mesh, 12-gauge welded wire is usually satisfactory for ducks of all ages over 3 wk.

Mesh size should be reduced to 1.9 cm for ducklings less than 3 wk of age. Vinyl-coated wire is preferable, but stainless steel or smooth, galvanized wire floors are satisfactory. Slats are not recommended for ducks because leg abnormalities have developed in ducks kept in research pens with slatted floors. Raised plastic flooring is commonly used in commercial duck production and is superior to wire in terms of reducing foot and hock damage.

Irritation to the feet and legs of ducks is reduced greatly if hard flooring such as wire occupies only a portion of the total floor area of a pen. In large floor pens, one-third wire and two-thirds litter is a satisfactory combination, provided that drinking devices are located on the wire-covered section of the pen, which greatly reduces the transport of water from the drinking area to the litter. Maintenance of litter in a satisfactorily dry condition is considerably more difficult in housing for ducks than for chickens and turkeys. Ducklings drink approximately 20 per cent more water than they need for normal growth and, as a result, the moisture content of their droppings is relatively high—approximately 90 per cent. To offset this extra water input in duck houses, extra litter and removal of excess water vapour by the ventilation system are essential. Supplemental heat may be necessary to aid in moisture control.

PERCHES

Egg-laying strains of chickens housed in cage-free systems are highly motivated to use perches at night. An entire flock will utilize perches at night if sufficient roosting space is provided.

Perches allow hens to roost comfortably with a minimum of disturbance and provide the opportunity for hens to seek refuge from aggressive birds so as to avoid cannibalistic pecking. Perches also minimize bird flightiness. Early exposure to perches during rearing encourages adult perching behaviour leading to a lower incidence of floor eggs. Adult Spanish breeds of chickens housed on a slatted/litter combination floor with perches compared with no perches were less stressed. However, if perches are not designed properly, they can lead to keel bone deformities. Perches should be designed to allow hens to wrap their toes around the perch and to balance themselves evenly on the perch in a relaxed posture for an extended period of time.

The perch should be elevated high enough from the surface floor to allow hens to grasp the perch without trapping their claws between the perch and the floor and to discourage the harbouring of mites. The center of the upper surface of the perch should be flat to allow for weight distribution so as to minimize keel deformities and foot problems. Perch edges should be smooth and round. The perch should be made of non-slip material. Ideally, perches should be positioned over slats or wire to prevent manure accumulation under the perches. Perch placement should minimize fecal contamination of birds, drinkers, and feeders below.

EGG-LAYING STRAINS

All hens should be able to roost at the same time; therefore, provide a minimum of 15 cm of usable linear perch space per egg-laying strain of chicken. Perforated floors that have perches incorporated into the floor structure and the rail in front of nest boxes can be counted as perch space.

A minimum of 20 per cent of the perch space should be elevated above the adjacent floor. Perches also need to be away from the wall at a sufficient distance to allow birds to use the perch. The height of the perch should not exceed 1 m above the floor so as to minimize skeletal fractures during bird flight from a perch. Provide enough space to allow a bird to jump down from its perch at an angle no steeper than 45°. Perches should be at least 30 cm apart to minimize cannibalistic pecking between birds on parallel roosts.

MEAT-TYPE CHICKENS

Only about 20 per cent of broilers in a flock will use perches at a single time. Depending on bird size, each broiler requires a perch space of 15 to 20 cm. If colony size is 100 birds and bird size indicates 20 cm of perch space/bird, then provide 400 cm of perch space/100 birds for 20 per cent usage. The width of the perch can range from 4 to 6 cm with perch heights of 10 to 30 cm depending on bird size. Broiler breeder hens prefer a roost with a width of 5 cm over narrower roosts of 3.8 cm and 2.5 cm. For adult broiler breeders, provide 28 cm of elevated roost per bird.

TURKEYS

If perches are to be used for turkeys, provide a minimum of 30 cm to 40 cm of elevated roost per bird. Perch height is dependent on bird size relative to breed, sex, and age of marketing with ranges from 20 to 150 cm. Turkeys appear to do well on wooden perches with rounded edges with dimensions of 5 cm in height and 7.5 cm in width.

NESTS

Hens place a high value on accessing nests, and their motivation for use increases greatly as the time of oviposition approaches. Hens without prior exposure to nests also show strong motivation to use nests for egg laying. Nests facilitate egg collection and minimize the risk of cloacal cannibalism. Because eggs laid in nests are cleaner and more sanitary, every effort should be made to avoid floor eggs. Use of electrical hot wire near walls outside of the nests may discourage the laying of floor eggs, as may a bright light that eliminates shadows when directed towards the corner. Pullets intended for systems with nests should be reared with access to raised areas and perches from an early age to become adept at moving up and down in space. Pullets allowed to access perches during rearing are less likely to lay eggs on the floor during the laying period.

Birds should be transferred to the layer house before sexual maturity to allow for sufficient time for exploration of the house and to find the nests before onset of lay. Nests should be dark inside.

Lights in nest boxes should be avoided because of increased risk of cannibalism. Nests should be constructed and maintained to protect hens from external parasites and disease organisms. Nests should be closed to bird access at night and re-opened before lay early in the morning. Nests should be regularly inspected and cleaned as necessary to ensure that there is no manure accumulation.

Nests should be provided with a suitable floor substrate that encourages nesting behaviour. Nests with wire floors or plastic-coated wire floors alone should be avoided. The provision of loose litter material in nests can be useful for training hens to use nests. For individual nest boxes with a single opening, provide a minimum of 1 nest box per 5 birds. Nest size for hens of egg-laying strains, which includes egg producers and layer breeders, can be 30 cm wide by 30 cm deep by 36 cm high.

Nests for broiler breeders are slightly larger than those for egg-laying strains of chickens with recommendations of 36 cm wide by 30 cm deep by 36 cm high. Turkey breeders require a nest size of 51 cm wide by 61 cm deep by 61 cm high, whereas duck breeders are provided a nest size of 36 cm wide by 45 cm deep and 30 cm high. For colony nests, provide a minimum of 0.8 m^2 of nest space per 100 chickens. Use of colony nests with duck breeders is not recommended because of increased incidence of floor eggs, egg breakage, and egg eating compared with individual nests. Hotter climates may require more nest space.

BROODING TEMPERATURES AND VENTILATION

Because thermoregulatory mechanisms are poorly developed in young chicks, poults, and ducklings, higher environmental temperatures are required during the brooding period. Requirements of young birds may be met by a variety of brooding environments.

Ventilation is ordinarily gradually increased over the first few weeks of the brooding period. Whether ventilation is by a mechanical system or involves natural airflow, drafts should be avoided, and streams of air that impinge upon portions of pens or groups of cages should be minimized. In relatively open brooding facilities, as in houses having windows for ventilation and with chicks kept in floor pens, draft shields may prove beneficial up to 10 d after hatching.

Young birds may huddle together or cluster when sleeping but are likely to disperse when awake. Within limits, birds can maintain appropriate body temperatures by moving away from or towards sources of heat when that is possible and by seeking or avoiding contact with other individuals.

Extreme huddling of young birds directly under the source of heat, especially during waking hours, usually indicates a need for more supplemental heat; dispersal associated with panting indicates that the environment is too warm. With brooding systems that allow birds to move towards or away from heat sources, the temperature surrounding the brooding area should be at least 20 to 25°C during the first few weeks but not be so high as to cause the young birds to pant or show other signs of hyperthermy.

When the entire room is heated and chicks are not free to move to cooler areas, the minimum temperatures that are recommended below may be too high. Thus, during the first week after hatching, a lower temperature may reduce the lethargy and non-responsiveness that is otherwise likely to be seen. Areas with minimum temperatures that are adequate for comfort and prevent chilling should be available to young birds.

The following minimum temperatures and weekly decreases are suggested until supplementary heat is no longer needed:

- For chicks, a 32 to 35°C ambient temperature initially, decreasing by 2.5°C weekly to 20°C; however, for some well-feathered strains, supplemental heat may be discontinued at 3 wk if room temperature is 22 to 24°C;
- For poults, 35 to 38°C, decreasing by 3°C weekly to 24°C;
- For ducklings, 26.5 to 29.5°C, decreasing by 3.3°C weekly to 13°C. After the brooding period, ducklings are comfortable at environmental temperatures of 18 to 20°C.

DUCKS

The recommended ventilation rates for chickens and turkeys have also given good results with ducks. Generally, however, lower relative humidity is desirable in duck houses to help offset the higher water content of duck droppings. Proper screening underneath watering equipment in houses with litter floors and the addition of generous amounts of litter are necessary features of the moisture control programme. When outside temperature allows, supplemental heat may be used to help to control moisture build-up in duck houses.

SEMEN COLLECTION AND ARTIFICIAL INSEMINATION

Semen collection and artificial insemination may be used in poultry depending on the species and type of research being conducted. Methods for semen collection and artificial insemination in poultry were developed in the

1930s and put into practice by the turkey industry such that artificial insemination is commonly used in commercial turkey breeding. Under conditions of artificial insemination, the breeder males and females are usually housed separately.

Careful and calm handling of the birds is needed to prevent injury and facilitates the success of the collections. Collection of semen from poultry involves restraining the male by the legs during the process. After stimulating the male by manual massage of the back area towards the tail, the semen is removed by squeezing the upper part of the cloaca and collected into a clean container. The number of cloacal strokes used should be limited to 4 strokes to avoid damage to the cloacal tissues. The semen may be inseminated without dilution or diluted with an extender. Males may be used for semen collection several times a week on alternate days although more than 3 collections per week may result in reduced semen volume and sperm concentration. The males must be acclimated to the handling and the semen collection process. Males may need to go through the procedure 3 to 4 times before they have a good response, but this can vary largely from male to male.

During the insemination process, the hen is gently restrained by the legs or held between the legs of the inseminator. Manual pressure is applied to evert the cloaca and expose the opening to the vagina. Semen is placed into the vaginal opening with an insemination straw, a small syringe or a pipette tip. Depth of insemination will vary with species.

As insemination occurs, the pressure on the cloaca is gradually released. After insemination, the hen should be gently released. If done correctly, the process takes only a few seconds to complete and should cause no pain or discomfort to the hen. Hens should also be acclimated to handling and the insemination process. If females are stressed or nervous, they may expel all or a portion of the semen immediately after the insemination. A typical insemination schedule that will give the highest level of fertility involves 3 inseminations within the first 10 d at the onset of reproduction, followed by insemination on a weekly basis. In turkey hens, morefrequent inseminations may be necessary to maintain fertility as they become older. Actual insemination schedules will vary depending on the research objectives.

STANDARD AGRICULTURAL PRACTICES

For handling birds and for all practices under this heading, experienced and skilled persons should carry out or train and supervise those who carry out these procedures.

BEAK TRIMMING

Trimming of the tip of the beak is done to minimize injury and death due to aggressive and cannibalistic behaviour. Outbreaks of cannibalism among egg-laying strains of chickens, turkeys, and ducks can occur with any housing system, resulting in a serious welfare problem. If the trimmed beak grows back, a second trim may be needed. An alternative to beak trimming is use of low light intensity in housing systems where light control is feasible. Genetic stock that shows little tendency towards cannibalistic behaviour and feather pecking should be used when possible.

EGG-STRAIN CHICKENS

Production, behaviour, and physiological measurements of stress and pain as indicated by neural transmission in the trimmed beak are used as criteria to determine well-being in beaktrimmed birds. In addition, the welfare of those hens that are pecked by beak-intact hens has been evaluated. Disadvantages of beak trimming include shortterm stress as well as short-term, and perhaps long-term, pain following the trimming of the beak. Because feeding behaviour must adapt to a new beak shape, a bird's efficiency in eating is impaired following a trim.

Welfare advantages include decreased mortality; reduced feather pulling, pecking, and cannibalism; better feather condition; less chronic stress; and less fearfulness and nervousness. Welfare advantages are more applicable to the interactive flock, whereas welfare disadvantages are applicable to individual birds whose beaks are trimmed. Genetic lines differ in their aggressiveness and beak-trimming requirements. Genetic selection is effective in reducing or eliminating most feather-pecking and beak-inflicted injuries and heritability estimates for survival suggest that the prospects for improving livability through genetic selection are good. Therefore, when feasible, stocks should be used that require either minimal or no beak trimming. Nevertheless, beak trimming is justified in stocks that otherwise are likely to suffer extensive feather-pecking and cannibalistic losses.

Management guides, available from most breeders, indicate methods for beak trimming to reduce these vices. Beak trimming should be carried out when birds are 10 d of age or younger. The amount of beak removed should be 50 per cent or less to avoid neuroma formation and to allow the keratinized tissue to regenerate. The length of the upper beak distal from the nostrils that remains following trimming should be 2 to 3 mm. The lower beak should be slightly longer than the upper beak. If a second trim is needed due to regrowth of the beak, it is recommended that it be done before the pullets are 8 wk of age to avoid a decrease in egg production.

BROILER-TYPE CHICKENS

Beak trimming is generally not required in young broilers raised for meat production. For broiler breeders, early beak trim before 10 d of age is generally sufficient to control feather-pecking and cannibalism in breeder stocks.

TURKEYS

Beak trimming of turkeys is a standard management practice. Strains of turkeys and sexes differ in their requirement for and their response to beak trimming. In strains of turkeys that exhibit a high incidence of beak-inflicted injuries, arc-type beak trimming at hatching is effective in reducing such injuries. Severe arc-type beak trimming increased mortality relative to hot-blade trimming of the upper beak at 11 d of age.

There was no evidence that arc-type beak trimming 1.5 mm from the nostrils at hatching or hot-blade trimming of the upper beak at 11 d of age increased mortality relative to leaving beaks intact. Beak trimming completed shortly after hatch did not modify performance or behaviour in commercial market toms compared with non-trimmed controls and also reduced pecking damage when beak regrowth did not occur. Arc-type beak trimming 1.5 mm anterior to the nostrils or hot-blade trimming of the upper beak at 11 d of age is recommended to prevent cannibalism in strains of turkeys that exhibit a high incidence of beak-inflicted injuries.

DUCKS

Feather pecking is a behaviour that sometimes occurs in ducks and may be controlled either by partial removal of the nail of the upper bill or inhibition of the growth of the nail by heat treatment. If not controlled, feather pecking injures the feather follicles of the tail, wings, and back, and the protective feather and down covering breaks down. Tip searing using cautery only may be a preferred method of bill trimming in Pekin ducks because of better weight gains following a trim and fewer changes in the morphology of the bill. For all species of poultry it is critical that the equipment used to trim beaks is maintained in good working condition. Personnel involved in beak trimming should receive species-specific training on proper procedures to use during beak trimming.

TOE TRIMMING

Because of the size and weight of the birds involved and the sharpness of their toenails, broiler breeder males and market turkeys generally have certain toes trimmed to prevent them from inflicting serious injuries to the hens during natural matings or to their penmates. Toe trimming should be done at 1 d of age using an electrical device that removes and cauterizes the third phalanx of the toes involved. Microwave energy application to the tip of the toe is also used to restrict toenail growth and is conducted using specialized equipment at the hatchery. In chickens, the microwave method did not result in increased stress or fearfulness. Provision of abrasive strips or hard surfaces in the facility may help to control excessive claw growth and reduce the need for declawing. Trimming toes for the purpose of identification is unjustified and should not be performed.

EGG-LAYING STRAINS OF CHICKENS

Leghorn hatchlings whose claws were trimmed through use of microwave energy experienced increased mortality and reduced feed consumption and BW during the pullet grow-out period. Removal of the claws resulted in a reduced foot spread allowing the toe of some pullets to slip into the wired mesh of the cage floor.

The pressure on the web between the toes led to a splitting of the foot epidermis in 24 of the 1,200 pullets whose claws were trimmed. Compton *et al.* reported similar results when using a hot blade to reduce claw length and suggested that chick movement about the wired cage was difficult until the toe grew long enough to allow the foot to spread across the wired cage floor. These results suggest that trimming the claws of egg-laying strains of chickens is not recommended.

BROILER BREEDER MALES

When meat-type males of certain genetic lines are to be used in natural matings, the practice of trimming certain toes at 1 d of age can be considered; toe trimming of breeding males may prevent injury to the female during natural mating. However, there is also evidence that toe trimming may impair the mating ability of males. The removal of one nail does not appear to cause chronic pain. For those genetic lines with long spurs, the spur bud on the back of the cockerel's leg may be removed at 1 d of age using a heated wire. Use of genetic lines with short, blunt spurs is preferable over spur removal. Most commercially available broiler breeder lines do not need to have their spurs removed.

TURKEYS

Toe trimming is a widespread management practice in turkey production. The number of toes trimmed per foot varies from 1 to 3 plus the dewclaw. Carcass grade of turkeys may or may not be improved by toe trimming, although rate of early mortality may be increased. Toe trimming may be justified when excessive injuries are likely to occur, but alternative methods should be considered to prevent bird injury.

SNOOD REMOVAL

Turkeys have a frontal process called a snood, which is an ornamental appendage for the adult male. The snood can be grasped by other turkeys during fighting and can be torn or damaged.

Breaks in the snood skin can be a health concern among older turkeys or those housed on pasture or on ranges. Data collected from industry showed that snood removal in tom poults reduced the odds of mortality. To avoid injury and possible infection, the snood can be removed from the newly hatched male poult by clipping or pinching the snood from its base on the head. If removed, the process should occur as soon as possible after hatching and no later than 3 wk of age. Snood removal after 3 wk of age is possible by clipping but not recommended without veterinary advice as the snood will continue to increase in size and vascularization especially in the males.

PARTIAL COMB AND WATTLE REMOVAL

Removal of part of the comb and wattles of chickens may be needed if birds are kept in cages. Combs and wattles can get caught in wire openings or feeders after significant comb and wattle growth has occurred. Comb and wattle removal is more commonly performed on cockerels because these structures are larger in males. Dubbing or removal of part of the wattles should only be used as a last resort when equipment or housing conditions cannot be modified to prevent torn or damaged combs or wattles. To perform successful comb and wattle removal with minimal bleeding and excellent long-term results, surgical scissors, scalpel blade, or electrocautery/radiosurgery electrode should be used to remove part of the comb and wattle during the first few days after hatching. To reduce risk of infection between birds, the scissor blades can be disinfected.

PINIONING

Surgical pinioning, which involves amputation of the wing tip from which primary feathers grow, or tendonotomy is used mainly in exhibit birds to render them permanently incapable of flight. Pinioning is not recommended as a means of reducing bird flightiness in chickens broilers, and ducks used for research and teaching. If flightiness is problematic, the primary feathers of one wing may be clipped.

INDUCED MOLTING

In birds, plumage is normally replaced before sexual maturity through a natural molt. Molting also occurs naturally after sexual maturity and is associated with a pause in egg production, which can be lengthy and take place out of synchrony with others in the flock. Inducing synchronized molting is used to rejuvenate laying flocks to extend the productive life of hens for 2 or 3 cycles of production.

Molting has become a common procedure for commercial table-egg layers and sometimes for broiler breeders and turkey breeders. In recycled egg-laying strains of chickens, molting decreases the demand for chicks by 47 per cent and thereby reduces the need to process, render, or bury the same percentage of spent hens. Rejuvenation of flocks also prevents the annual euthanasia of one hundred million additional male chicks. Additional advantages of molting include feather rejuvenation, thus improving thermoregulation. After a molt, livability and egg quality are improved during the second cycle of egg production compared with a non-molt control group.

EGG-STRAIN CHICKENS

Several procedures used to induce a molt have included short-term and long-term feed withdrawal; manipulation of dietary energy, protein levels, and dietary ingredients such as calcium, iodine, sodium, or zinc; and addition of feed additives that influence the neuroendocrine system such as iodinated casein. These procedures have been used coupled with a reduction in the daily photoperiod. These methods cause a cessation of egg production along with decreased BW and feather loss.

To allow for a return to egg laying, feather regrowth and BW gain are accomp-lished by feeding a diet designed to meet the nutritional requirements for a non-ovulating, feather-growing hen. Until 2000, the most common procedure used to induce a molt was to withdraw feed for 4 to 14 d without water restriction. Feed withdrawal for inducement of ovarian arrest is stressful leading to increased mortality during the first 2 wk of the molt. Hens are more fearful during a fasted molt compared with before and after a molt.

Temporary frustration as indicated by a moderate increase in aggression on the first day of feed removal has been noted in molted hens compared with non-molted full-fed controls. Aggression dissipated by the end of the first day, and molting hens showed elevated activity on the second day of fasting as indicated by increased non-nutritive pecking, standing, and head movement.

Resting behaviour increased by d 3 of fasting, and although non- nutritive pecking decreased from d 2, this pecking, interpreted as a redirection of foraging activity, remained higher than in control hens. Resting behaviour persisted for the remaining part of the fast. Similar changes in behaviour of hens subjected to a fasting molting regimen have been reported by Simonsen and Aggrey *et al.* with the notation of an additional behavioural repertoire of increased pre-ening on d 8 to 10 post-feed removal, most likely coinciding with the dropping of feathers. Hens subjected to a fasting molt compared with non-molted controls demonstrated decreased skeletal integrity, immunity, helper T cells and heterophil phagocytic activity. In addition, hens subjected to a fasting molt showed an increase in *Salmonella enteriditis* (SE) fecal shedding, the prevalence of SE in organs, inflammation of the intestines, the recurrence of a previous SE infection, and susceptibility to SE infection compared with non-molting controls.

Salmonella enteriditis was readily transmitted horizontally among molting birds under simulated field conditions, whereas in actual field settings, increased environmental *Salmonella* was observed in molted versus non-molted hens.

As an alternative to fasting, hens subjected to non-feed-removal molting regimens show post-molt performance not unlike the hens of the fasting molting regimen. Examples of successful non-feed-removal molting methods include the ad libitum feeding of diets high in corn gluten, wheat middlings, corn, or a combination of 71 per cent wheat middlings and 23 per cent corn. *Salmonella* shedding, intestinal inflammation, and internal organ contamination of SE-challenged hens were reduced and bone mineral density improved through the use of non-feed-withdrawal molting programmes compared with hens of a fasted molt.

Environmental presence of *Salmonella* increases during the molt in rooms containing fasting hens, but not in rooms of hens molted through wheat middlings. *Salmonella* fecal populations did not increase during a non-feed-removal molting programme compared with the pre-molt and post-molt periods, with *Salmonella* prevalence being the lowest during the molting period. Biggs *et al.* reported no differences in social behaviour between fasted hens and hens subjected to a non-feed-removal molting programme.

These results on increased resistance to *Salmonella* and improved skeletal integrity suggest that non-feed-withdrawal methods of molting should be used rather than the more conventional feed-withdrawal molting regimens. During the non-fast molt, hens should be monitored for health, mortality, and body weight. Water withdrawal or restriction, which can lead to increase mortality especially during hot weather, is not recommended.

BROILER BREEDERS, TURKEY BREEDERS, AND DUCK BREEDERS

Induced molt is occasionally done on parent breeding stock using feed withdrawal methods. Molting methods for breeder ducks are similar to those used for broiler breeders. Non-fasting methods of inducing a molt have not been reported in breeder stock.

SPECIAL CONSIDERATIONS

GENETICALLY MODIFIED BIRDS

To date, there are no special animal care requirements for transgenic or cloned poultry. Transgenic birds are cared for in the same manner as conventionally domesticated birds unless the genetic manipulation affects basic bird needs. Future transgenic animals may have special requirements and they should be cared for based on their genotype and phenotype rather than based on the technology that was used to create them.

SURGERIES

All intrathoracic and intraabdominal invasive surgeries require anesthesia. Caponization, or removal of the testes, is an invasive surgical procedure that requires anesthesia.

OTHER BIRD SPECIES

Gaunt and Oring and the Canadian Council on Animal Care offer recommendations on the care and use of wild birds, pigeons, doves, non-domesticated waterfowl, budgerigars, and quail. Parkhurst and Mountney provide animal care recommendations for geese, Coturnix quail, Bobwhite quail, chukar partridge, pheasants, guinea fowl, peafowl, pigeons, and swan. The Standing Committee of the European Convention for the Protection of Animals Kept for Farming Purposes provides recommendations and minimum standards for the welfare of ostrich and emu. Recommendations from New Zealand provide animal care guidelines for ratites. These references are given not as an endorsement but as referral material only.

AUTOIMMUNE DISEASES

The term "autoimmunity" literally means immunity against self and is caused by an immune-mediated reaction to self-antigens (*i.e.* failure of self-tolerance). Susceptibility to autoimmune disease has a genetic basis in humans and animals. Numerous viruses, bacteria, chemicals, toxins and drugs have been implicated as the triggering environmental agents in susceptible individuals.

This mechanism operates by a process of molecular mimicry and/or non-specific inflammation. The resultant autoimmune diseases reflect the sum of the genetic and environmental factors involved. Autoimmunity is most often mediated by T-cells or their dysfunction. As stated in a recent review "perhaps the biggest challenge in the future will be the search for the environmental events that trigger self-reactivity".

The four main causative factors of autoimmune disease have been stated to be:

1. Genetic predisposition;
2. Hormonal influences, especially of sex hormones;
3. Infections, especially of viruses; and
4. Stress.

IMMUNE-SUPPRESSANT VIRUSES

Immune-suppressant viruses of the retrovirus and parvovirus classes have recently been implicated as causes of bone marrow failure, immune-mediated blood diseases, hematologic malignancies (lymphoma and leukemia), dysregulation of humoral and cell-mediated immunity, organ failure (liver, kidney), and autoimmune endocrine disorders especially of the thyroid gland (thyroiditis), adrenal gland (Addison's disease), and pancreas (diabetes).

Viral disease and recent vaccination with single or combination modified live-virus vaccines, especially those containing distemper, adenovirus 1 or 2, and parvo virus are increasingly recognized contributors to immune-mediated blood disease, bone marrow failure, and organ dysfunction.

Genetic predisposition to these disorders in humans has been linked to the leucocyte antigen D-related gene locus of the major histocompatibility complex, and is likely to have parallel associations in domestic animals. Drugs associated with aggravating immune and blood disorders include the potentiated sulfonamides (trimethoprim-sulfa and ormetoprim-sulfa antibiotics), the newer combination or monthly heartworm preventives, and anticonvulsants, although any drug has the potential to cause side-effects in susceptible individuals.

Immune Deficiency Diseases

Immune deficiency diseases are a group of disorders in which normal host defences against disease are impaired. These include disruption of the body's mechanical barriers to invasion (*e.g.* normal bacterial flora; the eye and skin; respiratory tract cilia); defects in non-specific host defences (*e.g.* complement deficiency; functional white blood cell disorders), and defects in specific host defences (*e.g.* immunosuppression caused by pathogenic bacteria, viruses and parasites; combined immune deficiency; IgA deficiency; growth hormone deficiency).

IMMUNOLOGICAL EFFECTS OF VACCINES

Combining viral antigens, especially those of modified live virus (MLV) type which multiply in the host, elicits a stronger antigenic challenge to the animal. This is often viewed as desirable because a more potent immunogen presumably mounts a more effective and sustained immune response. However, it can also overwhelm the immunocompromised or even a healthy host that is continually bombarded with other environmental stimuli and has a genetic predisposition that promotes adverse response to viral challenge.

This scenario may have a significant effect on the recently weaned young puppy that is placed in a new environment. Today, given the recent major changes in vaccination guidelines for both dogs and cats, we are vaccinating less often and focusing on the core vaccines that all puppies and kittens should have (distemper virus, adenovirus, parvovirus, and rabies virus for dogs; panleukopenia virus, calicivirus, herpesvirus, and rabies virus for cats).

Vaccine Dosage

Manufacturers of combination (polyvalent) vaccines recommend using the same dose for animals of all ages and different sizes. It has never made any sense to vaccinate toy and giant breed puppies (to choose two extremes) with the same vaccine dosage. While these products provide sufficient excess of antigen for the average sized animal, it is likely to be either too much for the toy breeds or too little for the giant breeds. In addition, combining certain specific viral antigens such as distemper with adenovirus 2 (hepatitis) has been shown to influence the immune system by reducing lymphocyte numbers and responsiveness.

Hormonal State During Vaccination

Relatively little attention has been paid to the hormonal status of the patient at the time of vaccination. While veterinarians and vaccine manufacturers are aware of the general rule not to vaccinate animals during any period of illness, the same principle should apply to times of physiological hormonal change. This is particularly important because of the known role of hormonal change along with infectious agents in triggering autoimmune disease. Therefore, vaccinating animals at the beginning, during or immediately after an estrous cycle is unwise as would be vaccinating animals during pregnancy or lactation. In this latter situation, adverse effects can accrue not only to the dam but also because a newborn litter is exposed to shed vaccine virus. One can even question the wisdom of using MLV vaccines on adult animals in the same household because of exposure of the mother and her litter to shed virus.

Recent studies with MLV herpes virus vaccines in cattle have shown them to induce necrotic changes in the ovaries of heifers that were vaccinated during estrus. The vaccine strain of this virus was also isolated from control heifers that apparently became infected by sharing the same pasture with the vaccinates. Furthermore, vaccine strains of these viral agents are known to be causes of abortion and infertility following herd vaccination programmes. If one extrapolates these findings from cattle to the dog, the implications are obvious.

Killed Versus Modified Live Vaccines

Most single and combination canine vaccines available today are of MLV origin. This is based primarily on economic reasons and the fact that they produce more sustained protection. A long-standing question remains the comparative safety and efficacy of MLV versus killed (inactivated) virus vaccines, when a properly constituted killed vaccine is safer.

A recent examination of the risks posed by MLV vaccines concluded that they are intrinsically more hazardous than inactivated products. The residual virulence and environmental contamination resulting from the shedding of vaccine virus is a serious concern.

More importantly, the ability of new infective agents to develop and spread poses a threat to both wild and domestic animal populations. Vaccine manufacturers seek to achieve minimal virulence (infectivity) while retaining maximal immunogenicity (protection). This desired balance may be relatively easy to achieve in clinically normal, healthy animals but may be problematic for those with even minor immunologic deficit. The stress associated with weaning, transportation, surgery, subclinical illness and a new home can also compromise immune function. Furthermore, the common viral infections of dogs cause significant immunosuppression. Dogs harbouring latent viral infections may not be able to withstand the additional immunological challenge induced by vaccines. So, why are we causing disease by weakening the immune system with frequent use of combination vaccine products? After all vaccines are intended to protect against disease.

Recommendations

In response to questions posed above, veterinary vaccinologists have recommended new protocols for dogs and cats.

These include:

- Giving the puppy or kitten vaccine series followed by a booster at one year of age;
- Administering further boosters in a combination vaccine every three years or as split components alternating every other year until;
- The pet reaches geriatric age, at which time booster vaccination is likely to be unnecessary and may be unadvisable for those with aging or immunologic disorders.

In the intervening years between booster vaccinations, and in the case of geriatric pets, circulating humoral immunity can be evaluated by measuring serum vaccine antibody titers as an indication of the presence of "immune memory". Titers do not distinguish between immunity generated by vaccination and/or exposure to the disease, although the magnitude of immunity produced just by vaccination is usually lower.

Except where vaccination is required by law, all animals, but especially those dogs or close relatives that previously experienced an adverse reaction to vaccination can have serum antibody titers measured annually instead of revaccination. If adequate titers are found, the animal should not need revaccination until some future date.

Rechecking antibody titers can be performed annually, thereafter, or can be offered as an alternative to pet owners who prefer not to follow the conventional practice of annual boosters. Reliable serologic vaccine titering is available from several university and commercial laboratories and the cost is reasonable.

CANCER AND IMMUNITY

Proper regulation of cellular activity and metabolism is essential to normal body function. Cell division is a process under tight regulatory control. The essential difference between normal and tumor or cancerous cells is a loss of growth control over the process of cell division. This can result from various stimuli such as exposure to certain chemicals, viral infection, and mutations, which cause cells to escape from the constraints that normally regulate cell division. Proliferation of a cell or group of cells in an uncontrolled fashion eventually gives rise to a growing tumor or neo-plasm. Of course, tumors can be both benign (a localized mass that does not spread) or malignant (cancerous) in which the tumor grows and metastasizes to many distance sites via the blood or lymph. The situation in cancer is complex because not only can immunologically compromised individuals become more susceptible to the effects of cancer-producing viral agents and other chemical carcinogens, the cancer itself can be profoundly immunosuppressive. The form of immunosuppression usually varies with the tumor type. For example, lymphoid tumors (lymphomas and leukemia) tend to suppress antibody formation, whereas tumors of T-cell origin generally suppress cell-mediated immunity. In chemically induced tumors, immunosuppression is usually due to factors released from the tumor cells or associated tissues. The presence of actively growing tumor cells presents a severe protein drain on an individual which may also impair the immune response.

The body also contains a group of complimentary factors that provide a protective effect against tumors and other immunologic or inflammatory stresses. These are mixtures of proteins produced by T-cells and are referred to as "cytokines".

Cytokines include the interleukins, interferons, tumor-necrosis factors, and lymphocyte-derived growth factors. Recent studies have shown that normal levels of zinc are important to protect the body against the damaging effects of the specific cytokine, tumor-necrosis factor (TNF).

Currently about 15 per cent of human tumors are known to have viral causes or enhancement. Viruses also cause a number of tumors in animals and no doubt the number of viruses involved will increase as techniques to isolate them improve.

The T-cell leukemias of humans and animals are examples of those associated with retroviral infections. This same class of viruses has been associated with the production of autoimmunity and immunodeficiency diseases. The recent isolation of a retrovirus from a German Shepherd with Bcell leukemia exemplifies the potential role of these agents in producing leukemia and lymphomas in the dog.

The rising incidence of leukemia and lymphomas in an increasing number of dog breeds is a case in point. Similarly there has been an increase in the incidence of hemangiosarcomas (malignant tumors of the vascular endothelium) primarily in the spleen, but also in the heart, liver and skin. They occur most often in middle age or older dogs of medium to large breeds. The German Shepherd dog is the breed at highest risk, but other breeds including the Golden Retriever, Old English Sheepdog, Irish Setter and Vizsla have shown a significantly increased incidence especially in certain families. This suggests that genetic and environmental factors play a role. It is tempting to speculate that environmental factors that promote immune suppression or dysregulation contribute to failure of immune surveillance mechanisms. These protect the body against the infectious and environmental agents which induce carcinogenesis and neo-plastic change.

NUTRITIONAL APPROACHES FOR A HEALTHY IMMUNE SYSTEM

Wholesome nutrition is a key component for maintaining a healthy immune system and resistance to disease. Foodstuffs ingested by animals on a regular basis may be imbalanced in terms of major nutrients, minerals and vitamins, and often contain chemicals added to the final product to enhance its stability and shelf-life. Nutritional deficiencies or imbalances, as well as exposures to various chemicals, drugs and toxins, present a continual immunological challenge which can suppress overall immune function, especially in those animals genetically susceptible to immune dysfunction (immune deficiency, autoimmunity, allergies).

For example, Beagles are susceptible to distemper, demodectic mange, and autoimmune thyroid disease; Boxers to cancers; Doberman Pinschers and Rottweilers to parvovirus gastroenteritis with vomiting and diarrhea; and Shar Peis to IgA deficiency. However not all individuals at risk will become affected. One hypothesis to explain why some animals become affected with specific diseases while others remain normal, despite their common susceptibility, is called the "Threshold Model." In this situation, genetically susceptible individuals develop disease following the additive effects of inducing agents such as drugs, exposure to toxic or noxious substances, hormonal imbalances and dietary influences.

Genetic differences between individuals lead to quantitative variations in dietary requirements for energy, nutrients and overall health. Also genetic defects may result in inborn errors of metabolism that affect one or more pathways involving nutrients or their metabolites. Many inborn errors of metabolism are fatal, whereas others may show significant clinical improvement with nutritional management.

Minimal and maximal nutrient requirements that can be important in this regard include vitamin C, vitamin E and selenium, vitamin A, copper and vitamin B-12. Similarly, a wide variation occurs in the energy needs of dogs depending on their breed, age, sex and size. Breeders quickly learn to adjust the caloric intake of their animals depending on the optimal requirements of each individual.

Many environmental factors cause or trigger immune dysfunction leading either to immune deficiency or immune stimulation (reactive response, autoimmunity). One of the most common disorders of increasing prevalence today is autoimmune thyroid disease. Affected individuals have generalized metabolic imbalance and often have associated immunological dysfuncton. An important facet of managing these cases is minimizing exposure to unnecessary drugs, toxins and chemicals, and optimizing nutritional status with healthy balanced diets.

Because of the genetic predisposition to autoimmune disorders, the same recommendations apply to family members. Individuals susceptible to these disorders are at increased risk for adverse effects from immunological challenges following exposure to viruses, vaccines, and other infectious agents; a variety of chemicals, drugs and toxins and hormonal imbalances. Related to these events is the susceptibility to and development of cancer, a disruption of cell growth control.

NUTRITIONAL FACTORS AND THE IMMUNE SYSTEM

As alluded to above, an adequate nutritional state is important in managing a variety of inherited and other metabolic diseases as well as for a healthy immune system. Examples where nutritional management is important in inherited disorders include: adding ingredients to the diet to make it more alkaline for Miniature Schnauzers with calcium oxalate bladder or kidney stones; use of the vitamin A derivative, etretinate, in Cocker Spaniels and other breeds with idiopathic seborrhea of the skin; management with drugs and/or diet of diseases such as diabetes mellitus and the copper-storage disease prevalent in breeds like the Bedlington Terrier, West Highland White Terrier, and Doberman Pinscher; wheat-sensitive enteropathy in Irish Setters; and treatment of vitamin B-12 deficiency in Giant Schnauzers.

Other nutritional influences include the vitamin K dependent coagulation defect elicited in Devon Rex cats following vaccination; hip dysplasia in puppies fed excessive calories; osteochondritis dissecans (OCD) in dogs fed high levels of calcium; and hypercholesterolemia in inbred sled dogs fed high fat diets.

Nutritional factors that play an important role in immune function include zinc, selenium and vitamin E, vitamin B-6 (pyridoxine), and linoleic acid. Deficiency of these compounds impairs both circulating (humoral) as well as cell-mediated immunity. The requirement for essential nutrients increases during periods of rapid growth or reproduction and also may increase in geriatric individuals, because immune function and the bioavailability of these nutrients generally wanes with aging.

As with any nutrient, however, excessive supplementation can lead to significant clinical problems, many of which are similar to the respective deficiency states of these ingredients. Supplementation with vitamins and minerals should not be viewed as a substitute for feeding premium quality fresh and/or commercial dog foods.

NUTRITIONAL FACTORS AND THYROID METABOLISM

Nutritional influences can have a profound effect on thyroid metabolism. The classical example is the iodine deficiency that occurs in individuals eating cereal grain crops grown on iodine-deficient soil. This will impair thyroid metabolism because iodine is essential for formation of thyroid hormones. Another important link has recently been shown between selenium deficiency and hypothyroidism. Again, cereal grain crops grown on selenium-deficient soil will contain relatively low levels of selenium. While commercial pet food manufacturers compensate for variations in basal ingredients by adding vitamin and mineral supplements, it is difficult to determine optimum levels for so many different breeds of animals having varying genetic backgrounds and metabolic needs.

The selenium-thyroid connection has significant clinical relevance, because blood, but not tissue, levels of thyroid hormones rise in selenium deficiency. Thus, selenium-deficient individuals showing clinical signs of hypothyroidism could be overlooked on the basis that blood levels of thyroid hormones appear normal. The selenium issue is further complicated because synthetic antioxidants used to protect fats from rancidity can impair the bioavailability of vitamin A, vitamin E and selenium, and alter cellular membrane function and metabolism. As manufacturers of many premium cereal-based pet foods began adding the synthetic antioxidant, ethoxyquin, in the late 1980's, its effects, along with those of other synthetic preservatives (BHA or BHT), discussed below, are likely to be detrimental over the long term.

The total cumulative antioxidant load needs to be considered as well, because the common use of BHA to preserve animal fat sources is additive to the ethoxyquin incorporated into the finished product. An important question is what effect these induced changes in vitamin A, vitamin E and selenium quotients have wrought over the last 7-8 years with respect to the health and performance of companion animals. The way to avoid this potential problem is to use foods preserved with natural antioxidants such as vitamin E (tocopherol) and vitamin C (ascorbic acid) or prepare entirely home-cooked fresh natural ingredients.

SYNTHETIC AND NATURAL ANTIOXIDANTS

Synthetic antioxidants like BHA and BHT have been used as preservatives in human and animal foods for more than 30 years. Many pet food manufacturers prefer to use ethoxyquin today however, because of its excellent antioxidant qualities, high stability and reputed safety. But significant ongoing controversy surrounds issues related to its safety when chronically fed at permitted amounts in dog and cat foods. The same antioxidants have been linked to inducing or promoting a wide variety of cancers, although the published literature is both disturbing and contradictory in this regard. These safety questions pertain mostly to genetically susceptible breeds of inbred or closely linebred dogs. Toy breeds may be particularly at risk because they ingest proportionately more food and preservative for their size in order to sustain their metabolic needs.

Naturally occurring antioxidants (vitamin E and C) are also used in pet foods, and have become more popular in response to consumer and professional queries about the chronic effects of feeding synthetic chemical antioxidants to pets. While they are somewhat less effective and more expensive than the synthetic antioxidants, proponents of natural antioxidants believe their safety outweighs these drawbacks. It should be appreciated, however, that pet foods devoid of synthetic antioxidants added at the time of processing often contain ingredients (such as animal tallow or other fats and oils) that are preserved with antioxidants. Thus, claims made about the use of "all natural" antioxidant preservatives should also apply to preservatives used in the raw materials.

The net effect of these concerns has resulted in a major change in the pet food industry. Manufacturers of premium pet foods and most newly introduced foods have begun offering products preserved with natural antioxidants. This is a triumph for the views of the pet-owning public which prefers to use natural ingredients whenever possible.

NUTRITIONAL MANAGEMENT

Many veterinarians treating animals suffering from immunologic diseases appreciate that alternative nutritional management is an important step in minimizing their patients' environmental challenges. The results of this approach have been remarkable. Standard commercial diets containing synthetic chemical preservatives are replaced with naturally preserved foods. The replacement food must be of good quality and preferably of relatively low protein content (20-22 per cent). Increasing carbohydrate and reducing protein content, while maintaining high quality protein, has been shown to be beneficial for many affected animals and is also believed to have a positive effect on behaviour. Diet and behaviour appear to be linked because certain highly nutritious foods may contribute to deterioration in the condition of dogs with behavioural problems (dominant aggression, hyperactivity, and fear). For allergic animals, elimination diets are given for 6-8 weeks in order to evaluate their benefit to the patient. Homemade diets can be used instead of naturally preserved commercial diets, provided that the formula is properly balanced.

All other food supplements, including treats, are withdrawn, with the exception of such ingredients as fresh or stewed vegetables; low-fat cottage cheese or plain non-fat yogurt; boiled or scrambled eggs; chicken, turkey, fish, rabbit, venison or lamb stewed after removing the skin and fat; potatoes, steamed brown rice and pasta. A mixture of fresh vegetables, excluding onions and the cabbages, can be stewed together with basic kibbled cereal. Animals with known or suspected intolerance to dairy products, eggs or other ingredients should not be given these foods. A variety of individual diets can be developed by owners in consultation with their veterinarian and a nutritionist.

While animals treated in this way sometimes experience complete remission of symptoms, most show a 75 per cent or greater improvement of their existing condition. This holistic approach to diet is combined with the minimal amount of drug therapy required to extend the clinical remission and enhance the well-being and longevity of the patient. It is easy to implement, employs a common sense methodology and is inexpensive. The clients and patients like it, and best of all, it works!

NUTRITIONAL FACTORS AND THE IMMUNE FUNCTION

Nutritional factors that play an important role in immune function include zinc, selenium and vitamin E, vitamin B-6 (pyridoxine), and linoleic acid. Deficiencies of these compounds impairs both circulating (humoral) as well as cell-mediated immunity. The requirement for essential nutrients increases during periods of rapid growth or reproduction and also may increase in geriatric individuals, because immune function and the bioavailability of these nutrients generally wanes with aging.

As with any nutrient, however, excessive supplementation can lead to significant clinical problems, many of which are similar to the respective deficiency states of these ingredients. Supplementation with vitamins and minerals should only be given with the advice of a professional nutritionist and should not be viewed as a substitute for feeding premium quality fresh and/or commercial dog foods.

Bibliography

Ashok Ganguli : *Applied Biotechnology and Plant Genetics*, Oxford Book Company, Delhi, 2009.

Ashok Kumar Sharma : *Animal Biochemistry*, Random Publication, Delhi, 2012.

Asok Mukhopadhyay : *Animal Cell Technology*, IK Publication, Delhi, 2009.

B.D. Pandey and B.D. Joshi : *Bioresources for Rural Livelihood, Vol. I. Genetics*, Narendra Publication, Delhi, 2010.

C D N Singh : *Advanced Pathology and Treatment of Diseases of Domestic Animals*, International Book Distributing Co., 2008.

D. Gopalakrishna Rao : *A Text Book on Tumors of Domestic Animals*, International Book, Delhi, 2004.

J A Bierens De Haan : *Animal Behaviour,* Reprint Publication, Delhi, 2004.

James F. Crow and Motoo Kimura : *An Introduction to Population Genetics Theory*, Scientific Publication Delhi, 1998..

Lata Bhattacharya : *Animal Biochemistry*, Discovery Publication, Delhi, 2010.

Laurence M. Winters : *Animal Breeding*, Greenworld Publication, Delhi, 2002.

M. Sudhir : *Applied Biotechnology and Plant Genetics*, Dominant Publication, Delhi, 2000.

Madhusudan Mishra and Prabhakar Sharma : *A Textbook of Plant Genetics*, Wisdom Press, Delhi, 2012.

Mahipal Singh Rana : *An Introduction to Genetics* , Cyber Tech Publication, Delhi, 2009.

P. Kanakaraj : *A Text Book of Animal Genetics*, International Book Distributing Company, Delhi, 2007.

Prasum Tyagi : *A Textbook of Animal Physiology*, Dominant Publication, Delhi, 2010.

R S Chauhan and Kuldeep Dhama : *Aflatoxicosis in Animals and Its Public Health Significance*, International Book, Delhi, 2008.

R. Spier and J. Griffiths : *Animal Cell Biotechnology*, Academic Press, Delhi, 1998.

Ram Bramha Sanyal : *A Hand-Book of the Management of Animals in Captivity*, Natraj Publication, Delhi, 1998.

S. Nandi : *Animal Cell Culture and Virology*, New India Publication, Delhi, 2009.

S.K. Dwivedi and S. Gupta : *Agro-Animal Resources of Higher Himalayas*, Satish Serial Publication, Delhi, 2010.

S.K. Singh : *A Textbook of Molecular Genetics*, Campus Books International, Delhi, 2009.

S.P. Tiwari, S. Rajagopal and Usha Rani Mehra : *Analytical Techniques in Animal Nutrition*, Satish Serial Publication, Delhi, 2012.

Shambhu Dhir : *A Complete Book on Fish Genetics and Aquatic Environment*, Cyber Tech Publication, Delhi, 2010.

Shivangi Mathur : *Animal Cell and Tissue Culture*, Agrobios Publication, Delhi, 2006.

Swarnaprabha Pradhan : *Basics of Genetics*, Anmol Publication, Delhi, 2008.

T. Ramanathan : *Applied Genetics of Oilseed Crops*, Daya Publication, Delhi, 2004.

T.C. Jerdon : *A Natural History of all the Animals Known to Inhabit Indian Sub-continent*, Mittal Publication, Delhi, 2004.

T.N. Ananthakrishnan and K.G. Sivaramakrishnan : *Animal Biodiversity : Patterns and Processes*, Scientific Publication,

Delhi, 2006.

Udai Veer Singh : *Animal Biochemistry*, Sonali Publications, Delhi, 2011.

Vaibhav Suri : *A Modern Book on Biological Science—Cell/Genetics*, Cyber Tech Publication, Delhi, 2009.

William J.A. Payne and R. Trevor Wilson : *An Introduction to Animal Husbandry in the Tropics*, Wiley Publication, Delhi, 2013.

Yougesh Kumar and Rajeev Tyagi : *Aquaculture Fisheries Biotechnology and Genetics*, Manglam Publishers, Delhi, 2013

Index